PURIFYING AND CULTURING NEURAL CELLS

A LABORATORY MANUAL

ALSO FROM COLD SPRING HARBOR LABORATORY PRESS

RELATED TITLES

Addiction

Adult Neurogenesis (CSH Monograph Series 52)

The Biology of Alzheimer Disease

Drosophila *Neurobiology: A Laboratory Manual*

Imaging in Neuroscience: A Laboratory Manual

Molecular Cloning: A Laboratory Manual, Fourth Edition

Neuronal Guidance: The Biology of Brain Wiring

Parkinson's Disease

The Synapse

HANDBOOKS

At the Bench: A Laboratory Navigator, Updated Edition

At the Helm: Leading Your Laboratory, Second Edition

Experimental Design for Biologists

Lab Math: A Handbook of Measurements, Calculations, and Other Quantitative Skills for Use at the Bench

Lab Ref: A Handbook of Recipes, Reagents, and Other Reference Tools for Use at the Bench, Volume 1 and Volume 2

Statistics at the Bench: A Step-by-Step Handbook for Biologists

WEBSITE

www.cshprotocols.org

PURIFYING AND CULTURING NEURAL CELLS

A LABORATORY MANUAL

EDITED BY

Ben A. Barres
Stanford University School of Medicine

Beth Stevens
Boston Children's Hospital
Harvard Medical School

CSH PRESS
COLD SPRING HARBOR LABORATORY PRESS
Cold Spring Harbor, New York • www.cshlpress.org

PURIFYING AND CULTURING NEURAL CELLS
A LABORATORY MANUAL

Printed in the United States of America

Publisher	John Inglis
Acquisition Editor	Richard Sever
Director of Editorial Development	Jan Argentine
Managing Editor	Maria Smit
Developmental Editors	Heather Cerne, Tracy Kuhlman, Lori Martin
Project Manager	Maryliz Dickerson
Production Editor	Kathleen Bubbeo
Production Editor, *CSH Protocols*	Joanne McFadden
Production Manager	Denise Weiss
Director of Product Development & Marketing	Wayne Manos
Cover Designer	Ed Atkeson

Front Cover Artwork: Primary rat oligodendrocyte precursor cells (OPCs) were differentiated into oligodendrocytes for 3 d, and then fixed and stained for actin filaments (phalloidin, green) and the Arp2/3 complex (immunostaining, red), a key molecule that builds actin filaments in cells. Superresolution micrographs were obtained using a Zeiss Elyra Structured Illumination microscope. (Image credit: Brad Zuchero, Andrew Olson, and Ben Barres, Stanford Neurobiology.)

Library of Congress Cataloging-in-Publication Data

Purifying and culturing neural cells / [edited by] Ben A. Barres, Stanford University School of Medicine, Beth Stevens, Childrens' Hospital Boston, Harvard Medical School.

pages cm
Includes bibliographical references and index
ISBN 978-1-62182-011-6 (cloth) -- ISBN 978-1-936113-99-6 (pbk.)
1. Cell culture -- Laboratory manuals. 2. Neurons -- Physiology -- Laboratory manuals.
I. Barres, Ben, editor of compilation. II. Stevens, Beth, 1970- editor of compilation.

QP363.3.P87 2014
571.6'38--dc23

2013021663

10 9 8 7 6 5 4 3 2 1

Students and researchers using the procedures in this manual do so at their own risk. Cold Spring Harbor Laboratory makes no representations or warranties with respect to the material set forth in this manual and has no liability in connection with the use of these materials. All registered trademarks, trade names, and brand names mentioned in this book are the property of the respective owners. Readers should please consult individual manufacturers and other resources for current and specific product information.

With the exception of those suppliers listed in the text with their addresses, all suppliers mentioned in this manual can be found on the BioSupplyNet Web site at www.biosupplynet.com.

All World Wide Web addresses are accurate to the best of our knowledge at the time of printing.

Procedures for the humane treatment of animals must be observed at all times. Check with the local animal facility for guidelines.

Certain experimental procedures in this manual may be the subject of national or local legislation or agency restrictions. Users of this manual are responsible for obtaining the relevant permissions, certificates, or licenses in these cases. Neither the authors of this manual nor Cold Spring Harbor Laboratory assume any responsibility for failure of a user to do so.

The materials and methods in this manual may infringe the patent and proprietary rights of other individuals, companies or organizations. Users of this manual are responsible for obtaining any licenses necessary to use such materials and to practice such methods. COLD SPRING HARBOR LABORATORY MAKES NO WARRANTY OR REPRESENTATION THAT USE OF THE INFORMATION IN THIS MANUAL WILL NOT INFRINGE ANY PATENT OR OTHER PROPRIETARY RIGHT.

For a complete catalog of all Cold Spring Harbor Laboratory Press publications, visit our website at www.cshlpress.org.

Contents

CHAPTER 10

APPENDICES

APPENDIX 1

APPENDIX 2

General Safety and Hazardous Material Information

This manual should be used by laboratory personnel with experience in laboratory and chemical safety or students under the supervision of such trained personnel. The procedures, chemicals, and equipment referenced in this manual are hazardous and can cause serious injury unless performed, handled, and used with care and in a manner consistent with safe laboratory practices. Students and researchers using the procedures in this manual do so at their own risk. It is essential for your safety that you consult the appropriate Material Safety Data Sheets, the manufacturers' manuals accompanying equipment, and your institution's Environmental Health and Safety Office, as well as the General Safety and Hazardous Material Information in Appendix 1 for proper handling of hazardous materials in this manual. Cold Spring Harbor Laboratory makes no representations or warranties with respect to the material set forth in this manual and has no liability in connection with the use of these materials.

All registered trademarks, trade names, and brand names mentioned in this book are the property of the respective owners. Readers should please consult individual manufacturers and other resources for current and specific product information. Appropriate sources for obtaining safety information and general guidelines for laboratory safety are provided in General Safety and Hazardous Material Information (Appendix 2 of this manual).

Introduction

Ben A. Barres[1]

Stanford University School of Medicine, Stanford, California 94305-5125

The nervous system is the most complex organ in our bodies. In humans, it has been estimated to contain as many as one trillion cells, including neurons, glia, and vascular cells. Understanding the interactions between these cells and their networks is critical for understanding the development and function of our nervous system, as well as how these processes go awry in disease. How can we begin to understand and unravel such complexity? In other organs such as the immune system, purification of cell types has provided a powerful way forward. Beginning in the 1950s with the advent of immunostaining methods, immunologists developed antibodies for specific cell markers to detect epitopes of proteins, lipids, or carbohydrates that were specifically expressed by each immune cell. These antibody markers could then be used to identify, manipulate, and purify each immune cell type selectively, as well as to study the interactions of these cell types with each other in culture. It is almost impossible to conceive that the complexity of the immune system could have been unraveled without these methods.

In the late 1970s and 1980s, the immunologist Martin Raff, who in the early 1970s with Avrion Mitchison made important contributions to the development of antibody markers for specific immune cells, turned his attention to the nervous system. His laboratory elucidated many antibody markers for the surfaces of specific neural cell types, showing, for instance, that the anti-Thy1 monoclonal antibody recognized neurons, the anti-Ran2 monoclonal antibody recognized optic nerve astrocytes, the A2B5 antibody recognized oligodendrocyte precursor cells, and that anti-galactocerebroside (GC) antibodies recognized oligodendrocytes and Schwann cells (Brockes et al. 1977; Raff et al. 1978, 1979; Brockes et al. 1979; Bartlett et al. 1981; Abney et al. 1983; Miller et al. 1985; Williams et al. 1985; Raff 1989). Using these antibody markers, he was able to make important steps forward in understanding how these cells are generated during central nervous system development, particularly in the rat optic nerve, and was able to enrich various neural cell types using fluorescence-activated cell sorting (FACs) and other methods. This pioneering work was extremely influential and laid the groundwork for future studies of neural cell interactions.

As a Ph.D. student with Linda Chun and David Corey in the 1980s, I was able to use the Thy1 antibody to develop a method to purify retinal ganglion cells (RGCs) from the postnatal rat retina using immunopanning (Barres et al. 1988). In immunopanning, antibodies are adsorbed to the surface of a Petri dish and then cell suspensions are incubated on this dish to select for or deplete cell types of interest. In a typical immunopanning purification, a cell suspension from the retina or other nervous system region is prepared and then passed over several immunopanning dishes consecutively, with the first dish or two being used to deplete unwanted cell types such as microglia (which typically adhere via their Fc receptors to the first antibody-coated dish that they are placed on), and the final dish being used to select the cell type of interest. Although I could purify the RGCs to >99.5% purity, the purified cells quickly died by apoptosis after only a few hours in culture. This problem stymied me for several years. Fortunately, around that time, the peptide trophic factors BDNF and CNTF were identified.

---

[1]Correspondence: barres@stanford.edu

Cite this introduction as *Cold Spring Harb Protoc;* doi:10.1101/pdb.top070979

When I began my laboratory, we found that together these factors strongly promoted the survival of the purified RGCs in culture for weeks to months (Meyer-Franke et al. 1995).

Next, as a postdoctoral fellow with Martin Raff in the early 1990s, I was able to take advantage of the various antibody markers for glia that he and others had identified to develop simple immunopanning purification procedures for optic nerve astrocytes, oligodendrocyte precursor cells, and oligodendrocytes. Using these methods, we could isolate each of these cell types to >99.5% purity (Barres et al. 1992, 1993; Mi and Barres 1999). Once again, these highly purified cell types needed trophic factors to avoid apoptosis in culture, an important principle that so far appears to hold true for all purified cell types (Jacobson et al. 1997). Determining conditions that promote the survival of the majority of the purified cells can often be as hard, or even harder, than figuring out how to purify the cell type of interest. Alas, except for the optic nerve, Ran-2 did not turn out to be useful for purifying astrocytes from the central nervous system (CNS), and it was only after another 20 years of effort that we finally succeeded in developing a method to purify astrocytes by immunopanning and culture them, as is described later in this manual (Foo et al. 2011).

In my own laboratory over the past 20 years, we have continued to use immunopanning and sometimes FACS to develop purification and culture methods for many other CNS cell types, including other neural and glial types as well as vascular cells (Meyer-Franke et al. 1995; Hanson et al. 1998; Shi et al. 1998; Mi and Barres 1999; Goldberg et al. 2002; Cahoy et al. 2008; Dugas et al. 2008; Daneman et al. 2010a,b; Foo et al. 2011). The talented young people who have used and developed these methods, all of whom are now in their own laboratories, have each authored chapters detailing the purification and culture methods for these cell types. In nearly every case that we have purified and cultured a new CNS cell type, we have discovered something totally unexpected about its development or function. Purifying and culturing CNS cells has proved to be an incredibly powerful tool for dissecting fundamental neuron and glial properties, and especially powerful for understanding neuronal–glial interactions. For instance, by comparing the behavior of neurons cultured alone to those cultured with glia, we have been able to identify many fundamental roles for astrocytes that could never have been elucidated in a much more complex cellular mix.

This manual provides detailed protocols for purifying and culturing specific types of neurons, glia, and vascular cells from the central nervous system. First, we describe methods for retinal ganglion cells, corticospinal (upper) motor neurons, spinal (lower) motor neurons, and dorsal root ganglion sensory neurons. Next, we describe methods for astrocytes, pericytes and endothelial cells, and microglia. Last, we describe methods for oligodendrocyte precursor cells and oligodendrocytes, as well as for preparing myelinating cocultures of neurons and oligodendrocytes. We have not only provided step-by-step protocols, but also videos to demonstrate particularly critical steps in these procedures. We end with a troubleshooting chapter to discuss common pitfalls and errors.

By understanding the basics of each of the procedures described in this manual, you will have the tools to begin to invent your own procedures. There are many cell types in the nervous system that have never been purified. While the methods we present in this book have been developed to purify mouse and rat cell types, simple modifications should make it possible to purify and culture the counterpart cell types from humans or other species. In addition, emerging evidence suggests a profound degree of astrocyte heterogeneity (Tsai et al. 2012), and the methods described in this manual can easily be applied to isolate astrocytes from any CNS region desired.

It is interesting to compare this book with a previous manual for neural cell purification and culture titled *Culturing Nerve Cells*, which was edited by Gary Banker and Kimberly Goslin (1991, 1st edition; 1998, 2nd edition). Although now a bit dated, the purification and culture procedures described in that excellent book are still very useful. A limitation, however, of many of the original methods developed to purify specific neural cell types, such as astrocytes and Schwann cells, was that these methods were not prospective. They often relied on multiple purification steps that extended over several weeks in culture. This extended period of purification raised the concerns of subset selection or alteration of cell properties, so that it was not clear if the final purified cells were representative.

Cite this introduction as *Cold Spring Harb Protoc*; doi:10.1101/pdb.top070979

For instance, in the method developed by McCarthy and Vellis (1980) to purify astrocytes, a neonatal cell suspension was cultured in fetal calf serum for a week in the presence of a mitotic inhibitor drug such as cytosine arabinoside (AraC), and then after a few more days was "shaken" to remove a top layer of cells containing microglia and oligodendrocyte lineage cells. In these non-prospective isolation procedures, it was not clear what the starting yield of astrocytes was (as a percentage of astrocytes in the brain) and, if not a high yield, whether the selected cells were representative of astrocytes as a whole. Furthermore, as a result of the multiple steps extending over several weeks, there were many additional opportunities for selection of only a subset of astrocytes, perhaps not representative of most or all astrocytes. In fact, because these astrocytes could only be isolated from neonatal brain and not adult brain, it was possible that the surviving cells were some type of glial precursor cell rather than mature astrocytes. Thus, all of the methods described in this manual are prospective methods, in which the cell type of interest can be purified relatively acutely within a few hours. In each method, we have shown that the yield of cells isolated is at least 50%–100% of the number of cells of that type present in the starting brain cell suspension.

In addition, most of these previous procedures rely on the use of fetal calf serum to promote cell survival, proliferation, and growth, which is unphysiological not only because the blood–brain barrier normally prevents access of serum into the CNS, but because obviously fetal calf serum is not normally found in postnatal or adult rodents. Over the years, the presence of serum has been found to perturb the normal behavior of neural cells in ways that are quite concerning. For instance, as a Ph.D. student I noticed that astrocytes expressed different types of voltage-dependent calcium channels depending on what lot of fetal calf serum had been used for culture (Barres et al. 1989). The presence of serum in the culture medium induces astrocytes to express many genes normally only expressed by reactive astrocytes (Cahoy et al. 2008; Foo et al. 2011; Zamanian et al. 2012). Another example is Schwann cells, which when cultured in serum forever lose the ability to express myelin genes unless stimulated by pharmaceutical agents that elevate their intracellular levels of cAMP (Cheng and Mudge 1996). For this reason, in all of the methods described in this manual, the culture medium has been designed to be a defined serum-free medium—entirely lacking serum but usually containing one or more peptide trophic factors to promote cell survival, as well as amino acids and other important cell nutrients—so that it more closely resembles cerebrospinal fluid.

Of course, even when cells are prospectively purified and then placed into culture in a defined serum-free medium, it is possible that their properties may change over time so that they no longer are faithful replicas of their in vivo counterparts. This may happen because of subset selection: Even a given cell "type" such as retinal ganglion cells actually consists of as many as 20 subsets of different classes of RGCs. It is possible, even likely, that each class has dependence on different complements of survival signals (Masland 2001). Furthermore, the culture medium can never completely replicate faithfully the in vivo extracellular environment. Even the atmospheric oxygen is typically at 20% pO_2, rather than the 5% pO_2 level that occurs in the CNS in vivo. (Some researchers have found, for instance, that stem cells grow much better at 5% rather than 20% pO_2 [Csete 2005].) To address this, in some cases we have compared the gene expression pattern of acutely isolated cells of a given type to the expression pattern after a few days in culture. In the case of astrocytes, for example, we have found that the expression pattern of the acutely isolated, immunopanned astrocytes is nearly identical to their in vivo expression pattern (Cahoy et al. 2008). When we place those astrocytes in culture, however, after several days their expression pattern changes slightly, with about 50 genes being significantly different in level of expression (Foo et al. 2011; note that most of the genes that are different reflect the turning off of a few signaling pathways such as Notch, Wnt, or Shh, whose ligands are not present in our culture medium). In contrast, the gene expression patterns of astrocytes purified by the McCarthy–deVellis method differ by more than 700 genes, and the patterns suggest that these astrocytes are in fact more similar to glial precursors and/or reactive astrocytes (Foo et al. 2011; Zamanian et al. 2012). This is a powerful reminder of the importance of prospective isolation.

Two of the most powerful tools for prospectively purifying cells are immunopanning and FACS. Which is the better method? They can both be extremely useful. Over the years I have found that students in my laboratory invariably prefer to use immunopanning rather than FACS. Immunopanning is very gentle to the cells, whereas in FACS the cells are forced through a nozzle, which can be damaging to the typical, highly process-bearing neural cell. Immunopanning is also more rapid: Unless a laboratory owns a FACS sorter, time is needed on a core facility machine, which is also often distant from the laboratory where the purification is being performed. Last, the yields with immunopanning are often larger and more pure, because panning depletion steps are efficient and the adherent cells can be washed vigorously to remove nonadherent cells. In FACS, the cells must be sorted one by one and still must be double sorted to obtain good purity, both of which tend to decrease the final yield. That said, FACS is a very powerful tool for purification when combined with transgenic mice that express green fluorescent protein (or another fluorescent protein) in a cell type of interest.

Using the methods described in this manual, it is now possible to prepare serum-free cultures of prospectively isolated neurons and glia whose properties in culture very closely resemble their in vivo counterparts. This provides the opportunity to begin to do many types of studies to understand the fundamental properties of neurons and glia, as well as their interactions with each other and with vascular cells. So what sorts of studies can be done? With purified neural cell types, it is possible to study their properties immediately after isolation and before culture through gene profiling or western blotting. By simply acutely isolating different neural cell types and comparing their gene profiles, or by comparing the effects of coculture with astrocytes on neuronal gene expression, we have been able to quickly zoom in on genes that are crucial controllers of important biological processes; examples include the complement component C1q and the role of the classical complement cascade in helping to mediate developmental synapse elimination (Stevens et al. 2007; Stephan et al. 2012), and the master regulatory gene *MRF* in oligodendrocytes that controls oligodendrocyte differentiation and CNS myelination (Emery et al. 2009). We also identified that astrocytes were highly enriched in several different phagocytic pathways and therefore are likely to control synapse pruning by phagocytosis (Cahoy et al. 2008; WS Chung and BA Barres, in prep.).

Once neural cells are placed into culture, it is then possible to study the autonomous properties of neurons or glia cultured alone, and how they respond to specific environmental challenges such as activation of specific signaling pathways of interest. For instance, we were very surprised to find that purified retinal ganglion cells, when cultured alone, failed to survive in response to survival factors such as BDNF (even though they highly expressed the mRNAs for the BDNF receptor TrkB). When stimulated by depolarization, however, which elevated their cAMP levels, the cells demonstrated robust survival—at least in part because of a rapid recruitment of TrkB from an intracellular store to the RGC cell surface (Meyer-Franke et al. 1998; Goldberg and Barres 2000).

Purified neural cells can be used to study cell interactions such as neuronal–glial interactions. As examples, the use of purified cell types enabled us to discover that purified RGCs (and other neuron types that we tested), although they extended dendrites and axons and were electrically excitable, failed to form synapses unless astrocytes were present (Ullian et al. 2001; Eroglu and Barres 2010). The astrocytes not only strongly induced synapse formation but also strongly induced synaptic function (Pfrieger and Barres 1997; Allen et al. 2012). Moreover, by using cultures of purified neurons, we now had a culture assay to elucidate the molecular identity of the relevant astrocyte-secreted signals (which turned out to include thrombospondins, hevin, and glypicans [see Allen et al. 2012]). Similarly, we found that pericytes helped us to induce a mature blood–brain barrier by studying the interactions of pericytes with endothelial cells (Daneman et al. 2010b). As final examples, we identified a signaling pathway in oligodendrocytes, involving γ-secretase inhibition, that controls myelination (Watkins et al. 2008), and we found that oligodendrocytes secrete a protein that helps us to induce sodium channel clustering at nodes of Ranvier (Kaplan et al. 1997). This protein remains to be identified.

What else can be done with purified neural cells in culture? Neural cell types can be purified from mutant or transgenic mice, which can be an invaluable approach to study whether important cell–cell interactions such as synapse formation have been perturbed as a result of a given gene manipulation. Alternatively, viral vectors or RNA interference approaches can be used to manipulate gene expression of wild-type cells placed in culture. It is also possible to investigate what proteins are being synthesized or secreted and how signaling mechanisms regulate these processes. Metabolomics can now be used to

 Cite this introduction as *Cold Spring Harb Protoc*; doi:10.1101/pdb.top070979

study the identity of the small metabolites and other molecules that neurons, glia, or vascular cells secrete. The possibilities for experiments are literally endless!

Whatever experiments you do, it is always possible for artifacts to occur during culture. Before conclusions are drawn, it is crucial to determine the relevance of any in vitro findings by doing in vivo experiments to test their physiological significance. For instance, after identifying thrombospondins, hevin, and glypicans as astrocyte-secreted proteins that controlled the formation of excitatory synapses in vitro, we then examined mutant mice that lacked these proteins to see if synapse formation was similarly perturbed in vivo. (Fortunately, we found that it was; see Christopherson et al. 2005; Kucukdereli et al. 2011; Allen et al. 2012.) Such validation in vivo is always vitally important. When designing immunopanning experiments, it is also important to consider that it is a significant expense to purchase the required rodents, antibodies, and other panning and culture reagents. We believe that in most cases, the power of the many experiments enabled more than offsets this consideration.

This manual has been many years in the making. We held off on completing it while we finished and published our methods for purifying and culturing astrocytes. But there are many methods that we are still working on developing. What new chapters might be in the second edition of this manual? Hopefully it will soon be possible to purify and culture inhibitory neurons from the CNS. It would also be very useful to be able to purify and culture hippocampal neurons in serum-free medium, which at present is not possible as these neurons require the physical presence of astrocytes to survive in culture. Improved methods for the purification of microglia, which would allow immunopanning and exclusion of macrophages, are in the works. And, with any luck, we will be able to describe modifications to our current methods for purification and culture of human neurons and glia.

Whatever you are studying, we hope this manual will help you to explore the questions of interest to you in new ways.

REFERENCES

Abney ER, Williams BP, Raff MC. 1983. Tracing the development of oligodendrocytes from precursor cells using monoclonal antibodies, fluorescence-activated cell sorting, and cell culture. *Dev Biol* **100:** 166–171.

Allen NJ, Bennett ML, Foo LC, Wang GX, Chakraborty C, Smith SJ, Barres BA. 2012. Astrocyte glypicans 4 and 6 promote formation of excitatory synapses via GluA1 AMPA receptors. *Nature* **486:** 410–414.

Banker G, Goslin K. 1991. *Culturing neural cells.* MIT Press, Cambridge, MA.

Banker G, Goslin K. 1998. *Culturing neural cells*, 2nd ed. MIT Press, Cambridge, MA.

Barres BA, Silverstein BE, Corey DP, Chun LL. 1988. Immunological, morphological, and electrophysiological variation among retinal ganglion cells purified by panning. *Neuron* **1:** 791–803.

Barres BA, Chun LL, Corey DP. 1989. Calcium current in cortical astrocytes: Induction by cAMP and neurotransmitters and permissive effect of serum factors. *J Neurosci* **9:** 3169–3175.

Barres BA, Hart IK, Coles HS, Burne JF, Voyvodic JT, Richardson WD, Raff MC. 1992. Cell death and control of cell survival in the oligodendrocyte lineage. *Cell* **70:** 31–46.

Barres BA, Schmid R, Sendnter M, Raff MC. 1993. Multiple extracellular signals are required for long-term oligodendrocyte survival. *Development* **118:** 283–295.

Bartlett PF, Noble MD, Pruss RM, Raff MC, Rattray S, Williams CA. 1981. Rat neural antigen-2 (RAN-2): A cell surface antigen on astrocytes, ependymal cells, Müller cells and lepto-meninges defined by a monoclonal antibody. *Brain Res* **204:** 339–351.

Brockes JP, Fields KL, Raff MC. 1977. A surface antigenic marker for rat Schwann cells. *Nature* **266:** 364–366.

Brockes JP, Fields KL, Raff MC. 1979. Studies on cultured rat Schwann cells. I. Establishment of purified populations from cultures of peripheral nerve. *Brain Res* **165:** 105–118.

Cahoy JD, Emery B, Kaushal A, Foo LC, Zamanian JL, Christopherson KS, Xing Y, Lubischer JL, Krieg PA, Krupenko SA, et al. 2008. A transcriptome database for astrocytes, neurons, and oligodendrocytes: A new resource for understanding brain development and function. *J Neurosci* **28:** 264–278.

Cheng L, Mudge AW. 1996. Cultured Schwann cells constitutively express the myelin protein P0. *Neuron* **16:** 309–319.

Christopherson KS, Ullian EM, Stokes CC, Mullowney CE, Hell JW, Agah A, Lawler J, Mosher DF, Bornstein P, Barres BA. 2005. Thrombospondins are astrocyte-secreted proteins that promote CNS synaptogenesis. *Cell* **120:** 421–433.

Csete M. 2005. Oxygen in the cultivation of stem cells. *Ann NY Acad Sci* **1049:** 1–8.

Daneman R, Zhou L, Agalliu D, Cahoy JD, Kaushal A, Barres BA. 2010a. The mouse blood-brain barrier transcriptome: A new resource for understanding the development and function of brain endothelial cells. *PLoS One* **5:** e13741.

Daneman R, Zhou L, Kebede AA, Barres BA. 2010b. Pericytes are required for blood–brain barrier integrity during embryogenesis. *Nature* **468:** 562–566.

Dugas JC, Mandemakers W, Rogers M, Ibrahim A, Daneman R, Barres BA. 2008. A novel purification method for CNS projection neurons leads to the identification of brain vascular cells as a source of trophic support for corticospinal motor neurons. *J Neurosci* **28:** 8294–8305.

Emery B, Agalliu D, Cahoy JD, Watkins TA, Dugas JC, Mulinyawe SB, Ibrahim A, Ligon KL, Rowitch DH, Barres BA. 2009. Myelin gene regulatory factor is a critical transcriptional regulator required for CNS myelination. *Cell* **138:** 172–185.

Eroglu C, Barres BA. 2010. Regulation of synaptic connectivity by glia. *Nature* **468:** 223–231.

Foo LC, Allen NJ, Bushong EA, Ventura PB, Chung WS, Zhou L, Cahoy JD, Daneman R, Zong H, Ellisman MH, et al. 2011. Development of a method for the purification and culture of rodent astrocytes. *Neuron* **71:** 799–811.

Goldberg JL, Barres BA. 2000. The relationship between neuronal survival and regeneration. *Annu Rev Neurosci* **23:** 579–612.

Goldberg JL, Klassen MP, Hua Y, Barres BA. 2002. Amacrine-signaled loss of intrinsic axon growth ability by retinal ganglion cells. *Science* **296:** 1860–1864.

Hanson MG Jr, Shen S, Wiemelt AP, McMorris FA, Barres BA. 1998. Cyclic AMP elevation is sufficient to promote the survival of spinal motor neurons in vitro. *J Neurosci* **18:** 7361–7371.

Jacobson MD, Weil M, Raff MC. 1997. Programmed cell death in animal development. *Cell* **88:** 347–354.

Kaplan MR, Meyer-Franke A, Lambert S, Bennett V, Duncan ID, Levinson SR, Barres BA. 1997. Induction of sodium channel clustering by oligodendrocytes. *Nature* **386:** 724–728.

Kucukdereli H, Allen NJ, Lee AT, Feng A, Ozlu MI, Conatser LM, Chakraborty C, Workman G, Weaver M, Sage EH, et al. 2011. Control of excitatory CNS synaptogenesis by astrocyte-secreted proteins Hevin and SPARC. *Proc Natl Acad Sci* **108:** E440–E449.

Masland RH. 2001. The fundamental plan of the retina. *Nat Neurosci* **4:** 877–886. Review.

McCarthy KD, de Vellis J. 1980. Preparation of separate astroglial and oligodendroglial cell cultures from rat cerebral tissue. *J Cell Biol* **85:** 890–902.

Meyer-Franke A, Kaplan MR, Pfrieger FW, Barres BA. 1995. Characterization of the signaling interactions that promote the survival and growth of developing retinal ganglion cells in culture. *Neuron* **15:** 805–819.

Meyer-Franke A, Wilkinson GA, Kruttgen A, Hu M, Munro E, Hanson MG Jr, Reichardt LF, Barres BA. 1998. Depolarization and cAMP elevation rapidly recruit TrkB to the plasma membrane of CNS neurons. *Neuron* **21:** 681–693.

Mi H, Barres BA. 1999. Purification and characterization of astrocyte precursor cells in the developing rat optic nerve. *J Neurosci* **19:** 1049–1061.

Miller RH, David S, Patel R, Abney ER, Raff MC. 1985. A quantitative immunohistochemical study of macroglial cell development in the rat optic nerve: In vivo evidence for two distinct astrocyte lineages. *Dev Biol* **111:** 35–41.

Pfrieger FW, Barres BA. 1997. Synaptic efficacy enhanced by glial cells in vitro. *Science* **277:** 1684–1687.

Raff MC. 1989. Glial cell diversification in the rat optic nerve. *Science* **243:** 1450–1455.

Raff MC, Mirsky R, Fields KL, Lisak RP, Dorfman SH, Silberberg DH, Gregson NA, Leibowitz S, Kennedy MC. 1978. Galactocerebroside is a specific cell-surface antigenic marker for oligodendrocytes in culture. *Nature* **274:** 813–816.

Raff MC, Fields KL, Hakomori SI, Mirsky R, Pruss RM, Winter J. 1979. Cell-type-specific markers for distinguishing and studying neurons and the major classes of glial cells in culture. *Brain Res* **174:** 283–308.

Shi J, Marinovich A, Barres BA. 1998. Purification and characterization of adult oligodendrocyte precursor cells from the rat optic nerve. *J Neurosci* **18:** 4627–4636.

Stephan A, Barres BA, Stevens B. 2012. The complement system: An unexpected role in synaptic pruning during development and disease. *Annu Rev Neurosci*, in press.

Stevens B, Allen NJ, Vazquez LE, Howell GR, Christopherson KS, Nouri N, Micheva KD, Mehalow AK, Huberman AD, Stafford B, et al. 2007. The classical complement cascade mediates CNS synapse elimination. *Cell* **131:** 1164–1178.

Tsai HH, Li H, Fuentealba LC, Molofsky AV, Taveira-Marques R, Zhuang H, Tenney A, Murnen AT, Fancy SP, Merkle F, et al. 2012. Regional astrocyte allocation regulates CNS synaptogenesis and repair. *Science* **337:** 358–362.

Ullian EM, Sapperstein SK, Christopherson KS, Barres BA. 2001. Control of synapse number by glia. *Science* **291:** 657–661.

Watkins TA, Emery B, Mulinyawe S, Barres BA. 2008. Distinct stages of myelination regulated by γ-secretase and astrocytes in a rapidly myelinating CNS coculture system. *Neuron* **60:** 555–569.

Williams BP, Abney ER, Raff MC. 1985. Macroglial cell development in embryonic rat brain: Studies using monoclonal antibodies, fluorescence activated cell sorting, and cell culture. *Dev Biol* **112:** 126–134.

Zamanian JL, Xu L, Foo LC, Nouri N, Zhou L, Giffard RG, Barres BA. 2012. Genomic analysis of reactive astrogliosis. *J Neurosci* **32:** 6391–6410.

Cite this introduction as *Cold Spring Harb Protoc*; doi:10.1101/pdb.top070979

CHAPTER 1

Purification and Culture of Retinal Ganglion Cells

Alissa Winzeler and Jack T. Wang[1]

Department of Neurobiology, Stanford University School of Medicine, Stanford, California 94305

Retinal ganglion cells (RGCs) are the neurons that extend axons through the optic nerve, connecting and transmitting information from the retina to the brain. In mammals, RGCs receive information from bipolar and amacrine cells and synapse onto target cells in the lateral geniculate nucleus (LGN) as well as the superior colliculus. Methods for acute purification of RGCs from rodent retina by immunopanning followed by culture in a serum-free medium have facilitated the study of neuronal biology and function in a defined environment. These methods are introduced here, and modifications for achieving optimal RGC purity and culture are described.

RGC ANATOMY, FUNCTION, AND DEVELOPMENT

RGCs comprise a heterogeneous population of neurons in the retina collectively responsible for receiving, processing, and relaying visual information to the brain. A number of distinct classes of RGCs have been described based on morphology, electrophysiological properties, and central projection patterns, with distinct molecular markers identified for several RGC subtypes (Yamagata and Sanes 1995; Rockhill et al. 2002; Huberman et al. 2009; Kim et al. 2010; Huberman and Niell 2011; Masland 2012). The majority of RGC axons project into the contralateral hemisphere when the optic nerves cross at the optic chiasm, causing the visual system of each hemisphere to be dominated by information received by the contralateral eye. A subset of RGC axons, however, branch at the chiasm and send axons to the ipsilateral LGN and superior colliculus. Thus, each visual system target receives input from both eyes, allowing the signals to be integrated and compared. The spatial relationship between RGCs in the retina is recapitulated in the superior colliculus/tectum by opposing gradients of Eph/Ephrin signaling.

Among the earliest born cells in the retina, RGCs first appear between E10 and E14 in the rat retina; by E19, more than 95% have been generated (Reese and Colello 1992; Rapaport et al. 2004). The first RGC axons reach the superior colliculus by E16, whereas those of the latest-born RGCs arrive around P4-P5 (Dallimore et al. 2002). The growth rate of RGC nerve fibers is also dynamic during development, as recent work has demonstrated that perinatal rat RGCs undergo both a dramatic decrease in axon outgrowth speed and a concurrent increase in the extent of dendrite outgrowth around P0 because of retinal cues (Goldberg et al. 2002).

ADVANTAGES OF RGCS

RGCs have a number of advantages as a model system. First, visual input can be carefully controlled, and cellular responses to specific stimuli have been mapped in detail. Second, although the retina is

[1]Correspondence: jtw@stanford.edu

Cite this introduction as *Cold Spring Harb Protoc*; doi:10.1101/pdb.top070961

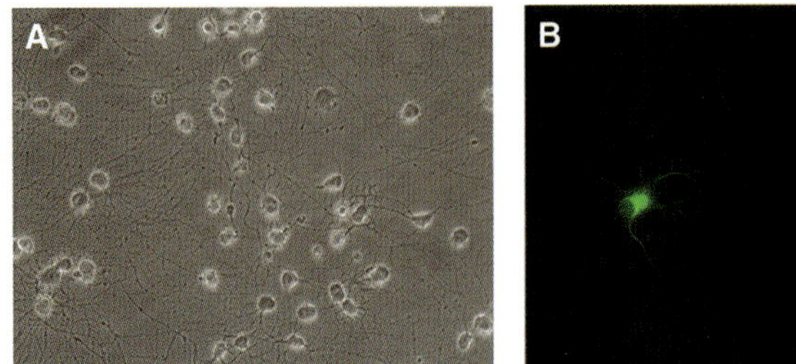

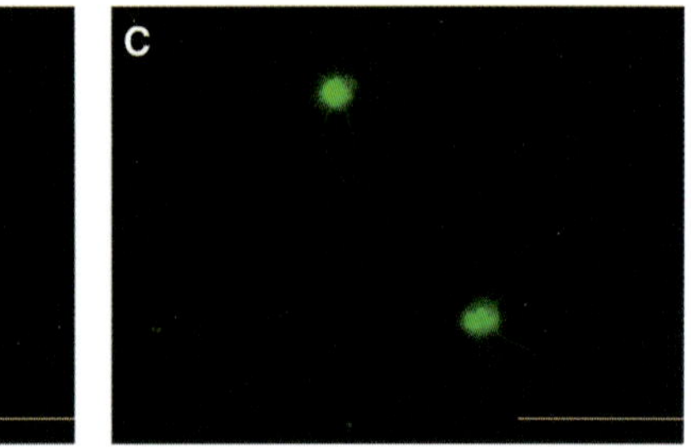

FIGURE 1. Purified RGCs cultured in growth medium. RGCs purified from postnatal day 7 rats were cultured on glass coverslips for 3 d in defined, serum-free medium. RGCs were plated at 10,000 cells per coverslip in a 24-well plate and imaged live on an inverted phase contrast microscope. Black bar, 50 µm (*A*). RGCs from the same sample were fixed and immunostained for the cytoskeletal protein tau (*B*) or PGP9.5 (*C*) to reveal processes. White bar, 25 µm.

part of the central nervous system (CNS), it is a relatively simple system comprised of a few well-characterized neuronal and glial cell types. As RGCs are CNS neurons with myelinated axons that come into contact with all of the major CNS cell types as they pass through the optic nerve, one can easily dissect the optic nerve away from the brain to study specific cell type interactions and functions in the neural retina. Indeed, protocols have been developed to isolate most of the neuronal and glial cell types from either the brain or the optic nerve.

Finally, RGCs have the advantage of being a well-established neuronal culture system (Fig. 1). They were the among the first CNS cell type to be isolated by panning, which allows for molecular identification of target cells without the harsh treatment required for fluorescence-activated cell sorting (FACS). Since then, much work has been done to determine the survival requirements of RGCs, and they can now be kept healthy for weeks in culture in the absence of either serum or other cell types. As the factors in serum are not well defined and can vary widely between sample batches, the presence of serum in culture has the potential to alter cellular responses dramatically. Thus, the ability to culture pure RGCs in defined, serum-free media has proven to be critical both in addressing how RGCs behave independently of other signals and in identifying the molecular signals that mediate specific neuronal–glial interactions in the CNS (Meyer-Franke et al. 1995; Ullian et al. 2001; Goldberg et al. 2002b; Christopherson et al. 2005; Watkins et al. 2008).

PRINCIPLES OF RGC ISOLATION AND CULTURE

RGCs were among the first CNS cell types to be purified by immunopanning (Barres et al. 1988). Twenty years later, the essentials of the procedure remain unchanged. Detailed methods are provided in Protocol 1: Purification and Culture of Retinal Ganglion Cells from Rodents (Winzeler and Wang). Briefly, retinas are dissociated into a single cell suspension after treatment with papain, and non-neuronal cell types are negatively selected using a combination of anti-macrophage antisera and/or *Bandeira simplifolica* lectin I (BSL-I), both of which deplete macrophages, microglia, and endothelial cells from the retinal cell suspension. RGCs are then pulled down through a positive panning process using an antibody recognizing the proteoglycan Thy-1 on the RGC cell surface. Note that two of the IgG anti-Thy1 monoclonal antibodies originally used by Barres et al. (1988), 2G12 and Ox-7, bind to Thy1 antigen with high affinity, which results in significant attachment to amacrine cells as well as RGCs. Thus, in rat, we switched to an anti-Thy1 IgM monoclonal antibody from the T11D7 hybridoma line, which is specific to but has a weaker affinity for Thy1. Because of this weaker affinity, the T11D7 antibody selects only for the highest Thy1-expressing neurons, the RGCs, which results in greater purity of the harvested cells. In mouse, an anti-Thy1.2 antibody is used to isolate murine RGCs.

To culture RGCs in the absence of serum, we previously identified a list of growth factors made by cells in contact with RGCs that promote RGC health and survival. These factors include BDNF

 Cite this introduction as *Cold Spring Harb Protoc*; doi:10.1101/pdb.top070961

(brain-derived neurotrophic factor), NT-4/5 (neurotrophin-4/5), CNTF (ciliary neurotrophic factor), LIF (leukemia inhibitory factor), insulin, IGF-1 (insulin-like growth factor 1), and bFGF (basic fibroblast growth factor) (Meyer-Franke et al. 1995). Furthermore, we found that these growth factors fall into three groups: BDNF and NT-4/5, CNTF and LIF, and insulin and IGF-1. We showed that factors from all three groups are required for optimal cell survival but factors within each group are redundant. Thus, adding factors from different groups produced additive effects on cell survival, while the addition of two factors from the same group promoted similar survival as either factor alone. In addition, we found that in the presence of BDNF, CNTF, insulin, and forskolin, bFGF is capable of promoting the survival of cells at clonal density but has no effect on cells at higher densities. Thus, the essential trophic factors in our culture media are comprised of BDNF, CNTF, and insulin, with bFGF only occasionally supplemented for low-density cultures.

We also showed that none of these factors is capable of promoting the survival of more than 2% of RGCs in culture without increasing intracellular cyclic AMP (cAMP) levels, either by directly depolarizing the cell or by adding forskolin, which activates adenylate cyclase to increase cAMP levels. Increased cAMP levels promote the recruitment of growth factor receptors such as TrkB to the cell surface from intracellular stores, thereby allowing the cells to become competent to respond to growth factor signaling (Meyer-Franke et al. 1998; Shen et al. 1999). Our finding is supported by a study showing that electrically stimulating RGCs cultured on silicon chips at a rate and pattern similar to that of endogenous activities in the developing retina increased the ability of trophic factors to promote RGC survival and neurite outgrowth in a cAMP-dependent manner (Goldberg et al. 2002a). Thus, the addition of forskolin to the culture medium mimics the effect of electrical activity in potentiating trophic factor response by about 10-fold, and is therefore a necessary component of RGC growth medium.

Finally, we found that several additional serum-free supplements further promoted optimal RGC survival. These include the Bottenstein–Sato supplement (SATO), thyroxine (T3) and a modified B27 serum-free supplement called NS21 (Chen et al. 2008). We found the original B-27 supplement (Invitrogen) to be toxic to RGCs after the manufacturer altered the way it was prepared. We have replaced it with NS21 with excellent results. The use of NS21 is also preferred because its composition is known precisely, whereas the composition of the B-27 supplement is proprietary and in our experience the quality of B-27 may vary dramatically from one lot to another.

Thus, the factors required for RGC culture include trophic factors (BDNF, CNTF, insulin), trophic potentiating agents (forskolin), and serum-free supplements (SATO, T3, NAC [*N*-acetyl-L-cysteine], glutamine, NS21) (for complete recipes and reagent information, refer to Protocol 1: Purification and Culture of Retinal Ganglion Cells from Rodents [Winzeler and Wang]). To date, the ability to isolate and culture RGCs by these means has been useful in a large variety of studies, including research on the molecular components of neuronal survival (Meyer-Franke et al. 1995; Shen et al. 1999; Goldberg et al. 2002a), neuronal–glial interactions that control synapse formation and function (Ullian et al. 2001; Christopherson et al. 2005; Eroglu et al. 2009; Allen et al. 2012), myelination and node of Ranvier formation (Kaplan et al. 1997; Watkins et al. 2008), and axon regenerative failure (Shen et al. 1999; Goldberg et al. 2002b).

REFERENCES

Allen NJ, Bennett ML, Foo LC, Wang GX, Chakraborty C, Smith SJ, Barres BA. 2012. Astrocyte glypicans 4 and 6 promote formation of excitatory synapses via GluA1 AMPA receptors. *Nature* **486:** 410–414.

Barres BA, Silverstein BE, Corey DP, Chun LL. 1988. Immunological, morphological, and electrophysiological variation among retinal ganglion cells purified by panning. *Neuron* **1:** 791–803.

Chen Y, Stevens B, Chang J, Milbrandt J, Barres BA, Hell JW. 2008. NS21: Re-defined and modified supplement B27 for neuronal cultures. *J Neurosci Methods* **171:** 239–247.

Christopherson KS, Ullian EM, Stokes CC, Mullowney CE, Hell JW, Agah A, Lawler J, Mosher DF, Bornstein P, Barres BA. 2005. Thrombospondins are astrocyte-secreted proteins that promote CNS synaptogenesis. *Cell* **120:** 421–433.

Dallimore EJ, Cui Q, Beazley LD, Harvey AR. 2002. Postnatal innervation of the rat superior colliculus by axons of late-born retinal ganglion cells. *Eur J Neurosci* **16:** 1295–1304.

Eroglu C, Allen NJ, Susman MW, O'Rourke NA, Park CY, Ozkan E, Chakraborty C, Mulinyawe SB, Annis DS, Huberman AD, et al. 2009. Gabapentin receptor α2δ-1 is a neuronal thrombospondin receptor responsible for excitatory CNS synaptogenesis. *Cell* **139:** 380–392.

Goldberg JL, Espinosa JS, Xu Y, Davidson N, Kovacs GT, Barres BA. 2002a. Retinal ganglion cells do not extend axons by default:

Promotion by neurotrophic signaling and electrical activity. *Neuron* **33:** 689–702.

Goldberg JL, Klassen MP, Hua Y, Barres BA. 2002b. Amacrine-signaled loss of intrinsic axon growth ability by retinal ganglion cells. *Science* **296:** 1860–1864.

Huberman AD, Niell CM. 2011. What can mice tell us about how vision works? *Trends Neurosci* **34:** 464–473.

Huberman AD, Wei W, Elstrott J, Stafford BK, Feller MB, Barres BA. 2009. Genetic identification of an On-Off direction-selective retinal ganglion cell subtype reveals a layer-specific subcortical map of posterior motion. *Neuron* **62:** 327–334.

Kaplan MR, Meyer-Franke A, Lambert S, Bennett V, Duncan ID, Levinson SR, Barres BA. 1997. Induction of sodium channel clustering by oligodendrocytes. *Nature* **386:** 724–728.

Kim IJ, Zhang Y, Meister M, Sanes JR. 2010. Laminar restriction of retinal ganglion cell dendrites and axons: Subtype-specific developmental patterns revealed with transgenic markers. *J Neurosci* **30:** 1452– 1462.

Masland RH. 2012. The neuronal organization of the retina. *Neuron* **76:** 266–280.

Meyer-Franke A, Kaplan MR, Pfrieger FW, Barres BA. 1995. Characterization of the signaling interactions that promote the survival and growth of developing retinal ganglion cells in culture. *Neuron* **15:** 805–819.

Meyer-Franke A, Wilkinson GA, Kruttgen A, Hu M, Munro E, Hanson MG Jr, Reichardt LF, Barres BA. 1998. Depolarization and cAMP elevation rapidly recruit TrkB to the plasma membrane of CNS neurons. *Neuron* **21:** 681–693.

Rapaport DH, Wong LL, Wood ED, Yasumura D, LaVail MM. 2004. Timing and topography of cell genesis in the rat retina. *J Comp Neurol* **474:** 304–324.

Reese BE, Colello RJ. 1992. Neurogenesis in the retinal ganglion cell layer of the rat. *Neuroscience* **46:** 419–429.

Rockhill RL, Daly FJ, MacNeil MA, Brown SP, Masland RH. 2002. The diversity of ganglion cells in a mammalian retina. *J Neurosci* **22:** 3831–3843.

Shen S, Wiemelt AP, McMorris FA, Barres BA. 1999. Retinal ganglion cells lose trophic responsiveness after axotomy. *Neuron* **23:** 285–295.

Ullian EM, Sapperstein SK, Christopherson KS, Barres BA. 2001. Control of synapse number by glia. *Science* **291:** 657–661.

Watkins TA, Emery B, Mulinyawe S, Barres BA. 2008. Distinct stages of myelination regulated by γ-secretase and astrocytes in a rapidly myelinating CNS coculture system. *Neuron* **60:** 555–569.

Winzeler A, Wang JT. 2013. Purification and culture of retinal ganglion cells from rodents. *Cold Spring Harb Protoc* doi: 10.1101/pdb.prot074906.

Yamagata M, Sanes JR. 1995. Target-independent diversification and target-specific projection of chemically defined retinal ganglion cell subsets. *Development* **121:** 3763–3776.

Cite this introduction as *Cold Spring Harb Protoc*; doi:10.1101/pdb.top070961

Protocol 1

Purification and Culture of Retinal Ganglion Cells from Rodents

Alissa Winzeler and Jack T. Wang[1]

Department of Neurobiology, Stanford University School of Medicine, Stanford, California 94305

Here we describe methods for acute purification of retinal ganglion cells (RGCs) from rodent retina by immunopanning, followed by culture in serum-free medium. Though the method was initially established and verified with rats, we have included modifications for the purification of mouse RGCs. This protocol is written for isolation of cells from one litter of pups. All of the volumes and numbers of panning plates should be scaled according to the number of litters used, particularly for rat RGCs.

MATERIALS

It is essential that you consult the appropriate Material Safety Data Sheets and your institution's Environmental Health and Safety Office for proper handling of equipment and hazardous material used in this protocol.

RECIPES: Please see the end of this protocol for recipes indicated by <R>. Additional recipes can be found online at http://cshprotocols.cshlp.org/site/recipes.

Reagents

BSA (4%)

To prepare a stock of 4% BSA in Dulbecco's phosphate-buffered saline (D-PBS), dissolve 8 g of BSA (Sigma-Aldrich A4161) in 150 mL of D-PBS (HyClone SH30264.01) at 37°C. Adjust the pH to 7.4 with ~1 mL of 1 N NaOH. Bring the volume to 200 mL. Filter through a 0.22-µm filter. Store in 1-mL aliquots at −20°C. (Dilute BSA stock to 0.2% in D-PBS just before use.)

DNase (0.4%)

To prepare a 0.4% stock of DNase in Earle's balanced salt solution (EBSS), add 1 mL of EBSS (Sigma-Aldrich E6267) per 12,500 units of DNase (Worthington LS002007). Keep on ice. Filter-sterilize, and store in 200-µL aliquots at −20°C.

Dulbecco's phosphate-buffered saline (D-PBS) with phenol red

Add 500 µL of 0.5% phenol red (Sigma-Aldrich P0290) per 500 mL bottle of Dulbecco's phosphate-buffered saline (D-PBS; HyClone SH30264.01).

Earle's Balanced Salt Solution (Ca^{2+}- and Mg^{2+}-free) (EBSS; Invitrogen 14155-063)

Ethanol-washed glass coverslips <R>

Fetal calf serum (FCS) (Gibco 10437-028)

Prepare 50-mL aliquots of FCS. Heat-inactivate aliquots for 30 min at 55°C, and then store at −20°C.

High-ovomucoid (high-ovo) stock solution (6×) <R>

[1]Correspondence: jtw@stanford.edu

Cite this protocol as *Cold Spring Harb Protoc*; doi:10.1101/pdb.prot074906

Immunopanning purification reagents

For purification of rat RGCs

Goat anti-mouse IgM, μ-chain specific (Jackson ImmunoResearch 115-005-020)
Goat anti-rabbit IgG (H + L) (Jackson ImmunoResearch 111-005-003)
Rabbit anti-rat macrophage polyclonal antibody (Cedarlane CLAD51240)
Thy1.1 hybridoma supernatant (see Protocol 2: Culturing Hybridoma Cell Lines for Monoclonal Antibody Production (Winzeler and Wang [a])

For purification of mouse RGCs

Goat anti-mouse IgG + IgM (H + L) (Jackson ImmunoResearch 115-005-044)
Lectin from *Bandeiraea simplicifolia* (BSL-1; Vector Labs L-1100) (5 mg/mL)
Mouse anti-mouse Thy1.2 (CD90) IgM (Serotec MCA02R)

Insulin stock (0.5 mg/mL) <R>
L-cysteine (Sigma-Aldrich C7477)
Laminin (mouse) (Cultrex; R&D Systems 3400-010-01)

Thaw mouse laminin (1 mg/mL) at 4°C. Make 10-μL aliquots and store at −80°C.

Low-ovomucoid (low-ovo) stock solution (10×) <R>
NaOH (1 N)
Neurobasal medium (Gibco 21103-049)
Papain (Worthington Biochemical LS003126)
Poly-D-lysine stock (PDL) (1 mg/mL; 100×)

Add 5 mL of H_2O to a 5-mg bottle of poly-D-lysine (PDL; Sigma-Aldrich P6407). Filter through a 2-μm filter. Make 100-μL aliquots and store at −20°C. Dilute stock to 1× in sterile H_2O before use.

Rat or mouse pups (P7)
RGC growth medium <R>

Warm growth medium to 37°C prior to cell culture.

Tris-HCl (50 mM, pH 9.5, sterile)

Dissolve 12.1 g of Trizma base in 200 mL of dH_2O. Adjust pH to 9.5 with HCl.

Trypan blue (Invitrogen 15250-061)
Trypsin stock (30,000 U/mL)

Dissolve trypsin (Sigma-Aldrich T9935) at 30,000 U/mL in EBSS. Filter through a 0.22-μm filter. Make 200-μL aliquots and store at −80°C.

Equipment

Centrifuge (tabletop, with 15- and 50-mL conical tube adaptors)
Dissection equipment

Dissection microscope
Forceps (#5 and #55; Fine Science Tools 11251-20 and 11255-20)
Retina spatula or probe to scoop retina (Fine Science Tools 10094-13)
Scalpel blade (#11) and scalpel blade handle (Fine Science Tools 10003-12)
Scissors (large, for decapitation) (ROBOZ RS-6820)
Scissors (small, curved) (e.g., ROBOZ RS-5603 for rat or Fine Science Tools 15011-12 for mouse)

Hemocytometer slide (Hausser Scientific 3110)
Nylon mesh filters (AmazonSupply 7050-1220-000-10)

Cut the nylon mesh into 3-inch squares, wrap in small packets of foil and autoclave.

Petri dishes (6-cm) (BD Falcon 351007)
Petri dishes (10- and 15-cm) (Falcon or Nunc)
Syringe filters (0.22-μm)

Cite this protocol as *Cold Spring Harb Protoc*; doi:10.1101/pdb.prot074906

Tissue culture incubator (37°C; 10% CO_2)
Tubes (conical) (15- and 50-mL)
Tubes (universal) (30-mL)
Water bath (37°C)

METHOD

An overview of the method is provided in Figure 1. It is a 2-d procedure; Steps 1–4 are performed on the first day.

Preparation

1. Add the secondary antibodies to the panning dishes as follows.

   For Rat RGCs

   i. Prepare two antibody-coated 15-cm Petri dishes for negative selection by adding 60 µL of goat anti-rabbit IgG (H + L) and 20 mL of 50 mM Tris-HCl (pH 9.5) per dish.
   ii. Prepare one antibody-coated 10-cm Petri dish for positive selection by adding 30 µL of goat anti-mouse IgM (µ-chain specific) and 10 mL of 50 mM Tris-HCl (pH 9.5).

   For Mouse RGCs

   iii. Prepare one antibody-coated 10-cm Petri dish for positive selection by adding 30 µL of goat anti-mouse IgG + IgM (H + L) and 10 mL of 50 mM Tris-HCl (pH 9.5).

2. Swirl the plates until the surfaces are evenly coated by the antibody-Tris solution.
3. Incubate the panning plates overnight at 4°C.

   Plates are hydrophobic to begin with, but after coating overnight, plates become visibly hydrophilic. If plates are needed immediately, a quick but not ideal way of making plates is to coat with secondary antibodies for 2 h at 37°C. For best results, however, it is preferable to coat the plates with secondary antibodies overnight.

4. Prepare the glass coverslips as follows.

   i. Place 15 to 20 ethanol-washed glass coverslips in a Petri dish.
   ii. Rinse the glass coverslips three times with sterile H_2O in the Petri dish.
   iii. After the last rinse, suction away any remaining H_2O and separate the coverslips such that they do not touch each other or the sides of the dish.
   iv. Allow the coverslips to completely air-dry.

      This should take 5–10 min after aspiration.

   v. Carefully add 100 µL of 1× PDL to the center of each coverslip.

      The PDL solution should cover the entire surface of the coverslip.

   vi. Incubate the coverslips for 30–45 min at room temperature.
   vii. Rinse the coverslips three times with sterile H_2O and aspirate until dry.
   viii. Dilute the mouse laminin (1 mg/mL) to a final concentration of 50 µg/mL by adding 10 µL of laminin stock to 5 mL of Neurobasal medium. Mix well and add 100 µL of the diluted laminin solution to each coverslip. Place in a 37°C incubator overnight.

5. Prepare three 15-mL conical tubes as follows.

   i. To one tube, add 10 mL of D-PBS and label as "Papain."
   ii. To a second tube, add 9 mL of D-PBS and label as "low-ovo."
   iii. To a third tube, add 5 mL of D-PBS and label as "high-ovo."

6. Add 10-15 mL of EBSS to a 10-cm Petri dish and warm in a 37°C incubator for at least 2 h.

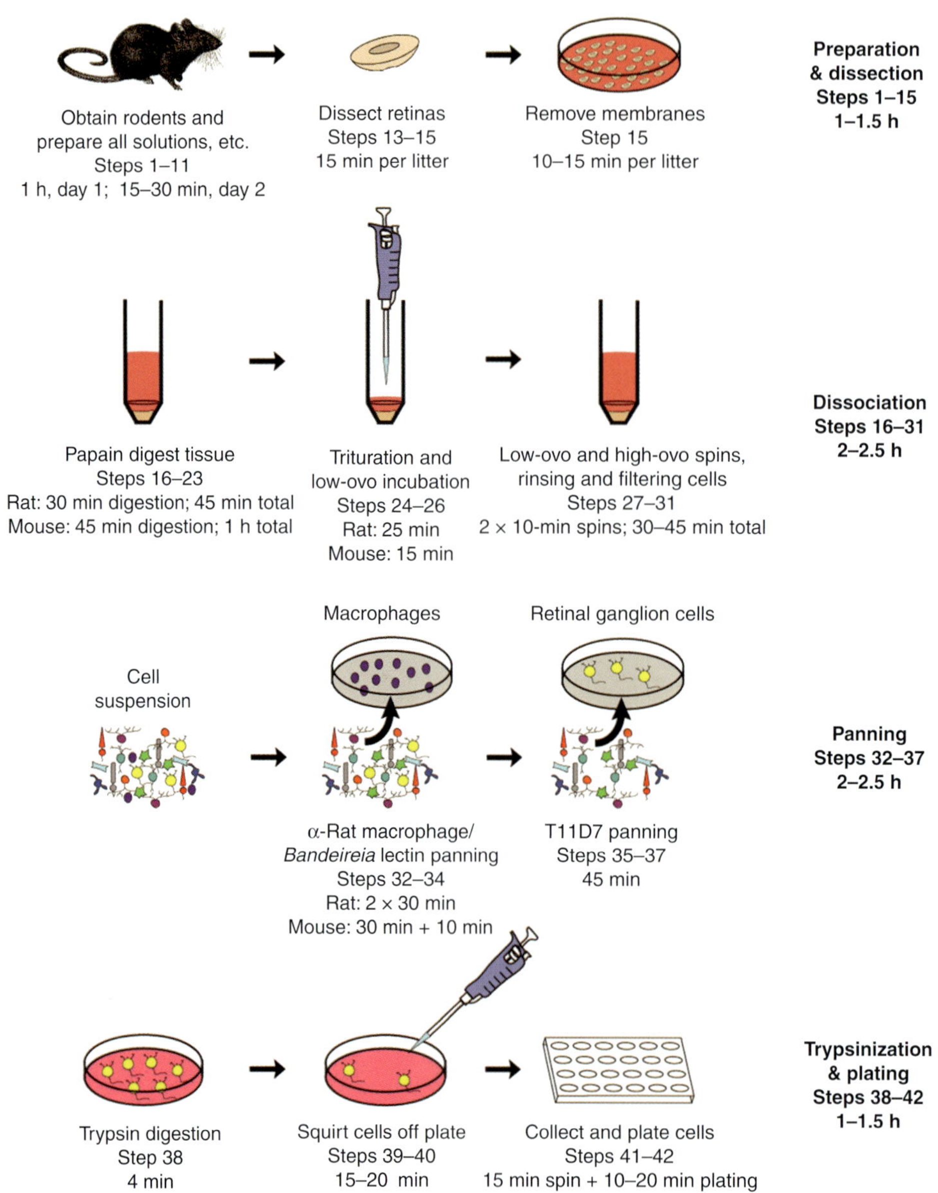

FIGURE 1. Purification of RGCs by immunopanning with anti-Thy1.

7. Prepare the panning dishes by coating with primary antibodies as follows.

For Rat RGCs

i. Rinse both 15-cm negative selection plates prepared in Step 1.i three times with D-PBS.

ii. Rinse the single 10-cm positive selection plate prepared in Step 1.ii with D-PBS. Add 10 mL of Thy1.1 hybridoma supernatant.

iii. Swirl to evenly coat plastic.

iv. Let plates sit at least 2 h at room temperature.

For Mouse RGCs

v. Make two new 15-cm plates for negative selection by adding 20 µL of BSL-1 and 20 mL of D-PBS to each plate for at least 2 h.

Cite this protocol as *Cold Spring Harb Protoc*; doi:10.1101/pdb.prot074906

vi. Rinse the 10-cm positive selection plate prepared in Step 1.iii three times with D-BPS. Add 10 µL of mouse Thy1.2 antibody to a mixture of 9 mL of D-PBS plus 1 mL of 0.2% BSA.

vii. Swirl to evenly coat plastic.

viii. Let plates sit at least 2 h at room temperature.

Bandeireia *lectin is an effective way to remove contaminating cell types. However, at high concentrations or when cells are incubated for long periods of time, it is also capable of binding to RGCs. Thus, care must be taken to check that the negative panning plates containing BSL-1 have not bound large numbers of RGCs. Although it is not in the protocol, BSL-1 can also be used as an alternative to anti-rat macrophage antibody for rat RGC preps. The steps are the same as those of the mouse preparation.*

The Thy1.2 antibody used for mouse RGC preps contains azide, which is toxic to cells. Thus, Thy1.2 plates must be washed thoroughly prior to exposure to cells (Step 34.ii).

8. Prepare the low-ovo and high-ovo solutions.

i. Add 2 µL of 1 N NaOH and 1 mL of low-ovo stock to the tube with 9 mL of D-PBS from Step 5.ii.

ii. Add 2 µL of 1 N NaOH and 1 mL of high-ovo stock to the tube with 5 mL of D-PBS from Step 5.iii.

Ovomucoid solutions are used to inhibit papain activity.

9. Prepare the panning buffer by combining 18 mL of D-PBS, 2 mL of 0.2% BSA, and 200 µL of insulin in a universal tube.

10. Prepare a 30% FCS solution by filtering 6 mL of FCS into 14 mL of D-PBS in a universal tube.

11. Prepare the RGC growth medium and warm in a 37°C incubator.

Dissection

The dissection process is critical to obtaining optimal yield. Retinas that are removed in many small pieces will be lost to the aspiration step prior to trituration. Thus, it is important to dissect out intact retinas to maximize yield. An illustration of Steps 13–15 is provided in online Movie 1 at cshprotocols.cshlp.org.

12. Euthanize a rat or mouse by CO_2 or direct decapitation.

13. Using a small scissor, cut away the skin covering the eyeball.

14. With the scalpel, make a horizontal slice across the exposed cornea starting at the lateral midpoint; the depth of the slit should be no more than 0.5 cm below the top of the cornea to avoid damaging the retina. Remove the lens with the back of the scalpel blade. Use the scalpel blade to gently scrape away the translucent vitreous humor that may appear after making the slit.

15. Holding the eye socket with forceps, use a flat instrument to gently scoop out the retina and place it in a 6-cm Petri dish containing D-PBS. Under a dissection microscope, use forceps to gently pull out and remove the membrane attached to the retina.

Often, blood vessels are visible coursing through the membrane.

16. Add 165 units of papain to the tube with 10 mL of D-PBS from Step 5.i. Shake the tube gently and warm in a 37°C water bath for 5 min or until the papain is dissolved.

17. Measure out 2 mg of L-cysteine and add it to the tube containing papain. Add ~10 µL of 1 N NaOH to neutralize the pH.

The addition of L-cysteine is necessary for the activation of the papain. Failure to add L-cysteine will decrease the efficacy of papain-mediated proteolysis.

The strength of papain solutions can vary widely from lot to lot and can affect the final yield. Although this protocol is based on the unit strength provided, each lot of papain should be tested empirically to determine optimal concentration.

18. Add 100 µL of DNase I stock to the tube containing papain.

19. Add 100 µL of DNase I stock to the tube containing low ovo solution prepared in Step 8.i.
20. Filter the papain solution into a universal tube.
21. Remove the excess D-PBS from the Petri dish containing the retina. Pour the retina into the tube containing filtered papain solution. Incubate in a water bath for 30 min (for rat retina) or 45 min (for mouse retina). Gently shake the tube to mix every 15 min.
22. Aspirate the papain supernatant and add 4 mL of the low-ovo solution to the retina. Let the tissue settle in the tube for 1 min.
23. For rat RGCs only, add 80 µL of anti-rat macrophage to the remaining 6 mL of low-ovo solution. (For mouse RGCs, proceed to Step 24.)

Trituration

An illustration of Steps 24 and 25 is provided in online Movie 2 at cshprotocols.cshlp.org.

24. Aspirate the low-ovo solution and add 2 mL of low-ovo + anti-macrophage solution (if rat) or low-ovo solution (if mouse) to the retina. Using a 1-mL pipette, gently triturate the retina three to four times, and then let the tissue settle for 1 min.
25. After 1 min, transfer the supernatant to a new 15-mL tube. Again add 1 mL of low-ovo + anti-macrophage or low-ovo solution to the tube containing the retina, gently triturate the retina three to four times, and let the tissue settle for 1 min. Repeat until all of the low-ovo solution has been used.

    Trituration is performed in low-ovo solution because cells survive trituration better in a lower protein solution. High-ovo is subsequently used to fully quench any residual papain enzymatic activity.

    Production of bubbles must be minimized, especially during trituration and recovery of trypsinized cells, to ensure optimal cell health.

26. For rat cells only: Incubate the retinas for an additional 10 min at room temperature to allow binding of the macrophage antibody.
27. Centrifuge the tissue at 1000 rpm for 12 min at 25°C.
28. For mouse cells only: During Step 27, rinse the two 15-cm plates prepared in Step 7.v with D-PBS and block each plate with 10 mL of 0.2% BSA.
29. Aspirate the supernatant and resuspend the cells in the high-ovo solution prepared in Step 8.ii. Centrifuge at 1000 rpm for 12 min at 25°C.

Panning

30. Resuspend the cells in panning buffer. Bring the volume up to 15 mL.
31. Filter the cell suspension through an autoclaved 20-µm nylon mesh.
    i. First, form a cone with the mesh on top of a 50-mL tube, then prewet the mesh with 1 mL of panning buffer.
    ii. Transfer the cell suspension through the mesh filter, 1 mL at a time.
    iii. Filter the remaining panning buffer through the mesh to rinse off any cells still on the mesh.

    Filtering the cell suspension through the nylon mesh gets rid of large debris and tissues, and helps break cell clumps into single cells, which will improve binding during the positive selection step (Step 35).

32. Transfer the cells to the first negative panning plate.
    i. For rat cells, pan for 20 min.
    ii. For mouse cells, pan for 30 min, shaking the plate every 15 min.

 Cite this protocol as *Cold Spring Harb Protoc*; doi:10.1101/pdb.prot074906

33. Transfer the cells to the second negative panning plate.

    i. For rat cells, pan for 45 min, shaking the plate every 15 min.

    ii. For mouse cells, pan for 10 min, or until the RGCs can be seen binding to the plate.

    We have found that mouse RGCs purified using this protocol may contain amacrine cells. If maximal purity of mouse RGCs is desired, such as for gene expression or proteomic studies, one can further deplete the amacrine population from the cell suspension by targeting VC1.1, an amacrine surface antigen (Goldberg et al. 2002). Transfer the cell suspension to an anti-VC1.1 antibody-coated negative dish (add 10 µL VC1.1 antibody to 30 mL of 50 mM Tris-HCl in a 10-cm dish) and pan for another 30 min (shaking the plate every 15 min) before proceeding to the positive panning step. This will help pull down additional amacrine cells.

    When panning, varying amounts of panning buffer can be used (10–20 mL/plate). Larger volumes will result in less clumping of cells within the cell suspension during panning, but have a tendency to spill once the cells are transferred to the smaller positive panning dishes.

34. Just before transferring the cells to the positive panning plate (Step 35), rinse the plate with D-PBS.

    i. For rat cells, discard the Thy1.1 hybridoma supernatant and rinse the plate three times with D-PBS.

    ii. For mouse cells, discard the Thy1.2 antibody solution and rinse the plate nine times with D-PBS (Thy1.2 antibody contains azide!).

35. Transfer the cells to the positive panning plate and pan for 45 min (for both rat and mouse RGCs), shaking the plate every 15 min.

Trypsinization

An illustration of Steps 39 and 40 is provided in online Movie 3 at cshprotocols.cshlp.org.

36. In the last 15 min of panning, add 100 µL of trypsin stock to 4 mL of warmed EBSS from Step 6.

37. Wash the positive panning dish six times with D-PBS. Examine the dish under the microscope to ensure only adherent cells remain.

38. Rinse the dish with the remaining EBSS from Step 6, then add the trypsin-EBSS solution. Incubate the dish at 37°C for 4 min.

    Ca^{2+} inhibits trypsin activity, so the plate must be rinsed with Ca^{2+}-free EBSS solution before the addition of trypsin.

    Optimal trypsinization time should be determined with each new batch of stock trypsin solution and should correspond to the point at which a large percentage of cells on the plate can be easily released by gentle pipetting. Trypsinization times that are either too short or too long can result in reduced cell viability.

39. Add 2 mL of 30% FCS to the panning dish. Gently squirt the solution against the plate to rinse off the adherent RGCs. Transfer the cell solution to a new 50-mL conical tube containing 1 mL of 30% FCS.

    Adding 1 mL of FCS to the tube before the cells are added ensures that the cells first contact liquid rather than a dry surface.

    Production of bubbles must be minimized, especially during trituration and recovery of trypsinized cells to ensure optimal cell health.

40. Add another 5 mL of FCS to the panning dish. Repeat squirting and transferring cells until all the FCS has been used. Check the dish under the microscope to confirm that the adherent cells have been washed off.

Plating

41. Remove a 50-µL aliquot of cell solution and quantify yield using a hemocytometer slide and trypan blue. Centrifuge the remaining cells at 1000 rpm for 12 min.

Expected yields are as follows:

- *~30,000 cells/P7 rat retina*
- *~15,000 cells/P7 mouse retina*

42. Resuspend the cells in prewarmed RGC growth medium and plate at the desired density on the PDL- and laminin-coated coverslips prepared in Step 4. Cell density can vary depending on desired experimental conditions: Plate <5000 cells per 24-well plate for low density, and >30,000 cells per 24-well plate for high density.

    To ensure optimal osmolarity and conditions for RGC survival and viability, cells are cultured in RGC growth medium, an enriched serum-free growth medium that contains 50% DMEM and 50% Neurobasal medium. For more information on rodent RGCs as a model system and details about their serum-free culture requirements, see Introduction: Purification and Culture of Retinal Ganglion Cells (Winzeler and Wang [b]).

RECIPES

BDNF Stock (50 µg/mL)

1. Prepare a master BDNF stock (1 mg/mL) by resuspending 1 mg of human brain-derived neurotrophic factor in powder form (BDNF; Peprotech 450-02) in 1 mL of cold, sterile 0.2% BSA (Sigma-Aldrich A-4161) that was prepared in D-PBS (Gibco 14287). Make 200-µL aliquots of the master stock, flash-freeze in liquid nitrogen, and store at −80°C.
2. To make a working BDNF stock, thaw a 200-µL aliquot of master stock on ice. At the same time, chill 3.8 mL of sterile 0.2% BSA solution on ice. Once chilled, add the master BDNF stock to the 0.2% BSA solution and mix well, but gently, to avoid foaming. Make 20-µL working aliquots and flash-freeze in liquid nitrogen. Store at −80°C.

Ciliary Neurotrophic Factor (10 µg/mL)

To prepare, dilute ciliary neurotrophic factor (CNTF; Peprotech 450-13) to 10 µg/mL in sterile 0.2% BSA that was prepared with D-PBS (Gibco 14287). Make 20-µL aliquots, flash-freeze in liquid nitrogen, and store at −80°C.

DMEM-SATO Base Growth Medium (with NB)

1. Combine the following:

Reagent	Amount (for 20 mL)	Final
Neurobasal medium (Gibco 21103-049)	9.5 mL	
DMEM (Gibco 11960-044)	9.5 mL	
Insulin stock (0.5 mg/mL) <R>	200 µL	5 µg/mL
Sodium pyruvate (100 mM; Gibco 11360-070)	200 µL	110 µg/mL
Penicillin-streptomycin (100×; Gibco 15140-122)	200 µL	100 U/mL (penicillin) 100 µg/mL (streptomycin)
SATO supplement (100×) <R>	200 µL	1×
Thyroxine (T3) stock (4 µg/mL) <R>	200 µL	40 ng/mL
L-Glutamine (200 mM; Gibco 25030-081)	200 µL	292 µg/mL
NS21 supplement (50×) (R&D Systems AR008; see Chen et al. 2008)	400 µL	1×
NAC stock (5 mg/mL) <R>	20 µL	5 µg/mL

2. Filter through a rinsed 0.22-µm filter to sterilize.

 Cite this protocol as *Cold Spring Harb Protoc*; doi:10.1101/pdb.prot074906

Ethanol-Washed Glass Coverslips

Extensively wash 12-mm glass coverslips (Carolina Biological Supply 633029) in 70% ethanol. Perform the washes on a platform shaker in a beaker, with enough motion to lightly agitate the coverslips but not break too many. Wash the coverslips for about 1 mo, exchanging the ethanol approximately every day. (It is fine to skip some exchanges.) Store the washed coverslips in 70% ethanol until use.

Forskolin Stock (4.2 mg/mL)

To prepare, add 1 mL of sterile DMSO to a 50-mg bottle of forskolin (Sigma-Aldrich F6886) and pipette up and down until the powder is fully resuspended. Transfer to a 15-mL conical tube and add an additional 11 mL of DMSO to achieve a final concentration of 4.2 mg/mL. Store in 20- and 80-μL aliquots at −20°C.

High-Ovomucoid Stock (6×)

1. Dissolve the following in 160–180 mL of Dulbecco's phosphate-buffered saline (D-PBS; Invitrogen 14287-080).

Reagent	Amount (for 200 mL)	Final concentration
BSA (Sigma-Aldrich A-8806)	6 g	30 mg/mL
Trypsin inhibitor (Worthington LS003086)	6 g	30 mg/mL

2. Adjust the pH to 7.4 with 10 N NaOH. Bring the volume to 200 mL with D-PBS, and then filter-sterilize.
3. Make 1-mL aliquots and store them at −20°C.

Insulin Stock (0.5 mg/mL)

To 20 mL of sterile water, add 10 mg of insulin (Sigma-Aldrich I6634) and 100 μL of 1.0 N HCl. Mix well. Filter through a 0.22-μm filter. Store at 4°C for 4–6 wk.

Low-Ovomucoid Stock Solution (10×)

To prepare, add 3 g of BSA (Sigma-Aldrich A8806) to 150 mL D-PBS. Mix well. Add 3 g of trypsin inhibitor (Worthington LS003086) and mix to dissolve. Add ~1 mL of 1 N NaOH to adjust the pH to 7.4. Bring the volume to 200 mL with D-PBS. Filter-sterilize through a 0.22-μm filter. Make 1.0-mL aliquots and store at −20°C.

NAC Stock (5 mg/mL)

To prepare, dissolve 50 mg of *N*-acetyl-L-cysteine (NAC) powder (Sigma-Aldrich A8199) in 10 mL of Neurobasal Medium (Gibco/Life Technologies 21103). (The solution will be yellowish.) Filter through a 0.22-μm filter. Prepare 20- and 80-μL aliquots and store them at −20°C.

RGC Growth Medium

DMEM-SATO base growth medium (with NB) <R>	20 mL
Forskolin stock (4.2 mg/mL) <R>	20 μL
BDNF stock (50 μg/mL) <R>	20 μL
Ciliary neurotrophic factor (10 μg/mL) <R>	20 μL

Warm growth medium to 37°C prior to cell culture, or store at 4°C for up to 3 d.

SATO Supplement (100×)

1. Prepare the following stock solutions (these should be made fresh; do not reuse).
   - Combine 5 mg of progesterone (Sigma-Aldrich P8783) and 200 µL of ethanol to make a progesterone stock solution.
   - Combine 4 mg of sodium selenite (Sigma-Aldrich S5261), 10 µL of 1 N NaOH, and 10 mL of Dulbecco's modified Eagle's medium (DMEM; Gibco/Life Technologies 11960-044) to make a sodium selenite stock solution.

2. Combine the following:

Reagent	Quantity (for 200 mL)	Final concentration (100×)
BSA (Sigma-Aldrich A4161)	2 g	10 mg/mL
Transferrin (Sigma-Aldrich T1147)	2 g	10 mg/mL
Putrescine (Sigma-Aldrich P5780)	320 mg	1.6 mg/mL
Progesterone stock solution	50 µL	6 µg/mL
Sodium selenite stock solution	2 mL	4 µg/mL

3. Bring to a total volume of 200 mL in DMEM, and then filter-sterilize. Aliquot and store at −20°C.

Thyroxine (T3) Stock (4 µg/mL)

Dissolve 3.2 mg of 3,3′,5-triiodo-L-thyronine sodium salt (T3; Sigma-Aldrich T6397) in 400 µL of 0.1 N NaOH. Add 10 µL of T3 solution to 20 mL of Dulbecco's phosphate-buffered saline (D-PBS; Gibco 14287). Filter through a 0.22-µm filter, discarding the first 10 mL. Make 200-µL aliquots and store at −20°C.

ACKNOWLEDGMENTS

We thank Maria Fabian for technical assistance, and Mariko Howe, Amanda Brosius-Lutz, and Jennifer Zamanian for reading and comments on the protocol.

REFERENCES

Chen Y, Stevens B, Chang J, Milbrandt J, Barres BA, Hell JW. 2008. NS21: Re-defined and modified supplement B27 for neuronal cultures. *J Neurosci Methods* **171:** 239–247.

Goldberg JL, Klassen MP, Hua Y, Barres BA. 2002. Amacrine-signaled loss of intrinsic axon growth ability by retinal ganglion cells. *Science* **296:** 1860–1864.

Winzeler A, Wang JT. 2013a. Culturing hybridoma cell lines for monoclonal antibody production. *Cold Spring Harb Protoc* doi: 101101/pdb.prot074914.

Winzeler A, Wang JT. 2013b. Purification and culture of retinal ganglion cells. *Cold Spring Harb Protoc* doi: 101101/pdb.prot070961.

Cite this protocol as *Cold Spring Harb Protoc*; doi:10.1101/pdb.prot074906

Protocol 2

Culturing Hybridoma Cell Lines for Monoclonal Antibody Production

Alissa Winzeler and Jack T. Wang[1]

Department of Neurobiology, Stanford University School of Medicine, Stanford, California 94305

This protocol describes how to culture hybridoma cell lines (e.g., Thy1.1) for monoclonal antibody production. Supernatants harvested from such cultures can be used to purify various rodent neural cell types by immunopanning.

MATERIALS

It is essential that you consult the appropriate Material Safety Data Sheets and your institution's Environmental Health and Safety Office for proper handling of equipment and hazardous material used in this protocol.

RECIPES: Please see the end of this protocol for recipes indicated by <R>. Additional recipes can be found online at http://cshprotocols.cshlp.org/site/recipes.

Reagents

Dimethyl sulfoxide (DMSO)

Fetal calf serum (FCS) (Gibco 10437-028)

Make 50-mL aliquots and heat inactivate for 30 min at 55°C, then store at −20°C.

Hybridoma cell line medium <R>

Hybridoma cells of interest (e.g., Thy1.1)

Isopropanol

RPMI medium (Invitrogen 21870-100)

Tris buffer (1 M, pH 8)

Equipment

Bottles (1 L, sterile)

Centrifuge (tabletop)

Cryojars (Nalgene)

Cryovials

Hemocytometer slide

Tissue culture flasks (75 cm^2, 150 cm^2)

[1]Correspondence: jtw@stanford.edu

Cite this protocol as *Cold Spring Harb Protoc*; doi:10.1101/pdb.prot074914

Tissue culture incubator (37°C, 10% CO_2)
Tissue culture plates (24 well)
Tubes (15 and 50 mL)
Water bath (37°C)

METHOD

1. At least 1 h before beginning culture, add 0.5 mL of hybridoma medium to each of the middle eight wells of a 24-well plate. Equilibrate in an incubator at 37°C with 10% CO_2 for at least 1 h. In addition, warm and equilibrate 12 mL of medium.
2. Remove the hybridoma cells from liquid nitrogen and quick-thaw in a 37°C water bath. Mix the cells into 10 mL of the prewarmed medium. Centrifuge the cells at 1000 rpm for 10 min.
3. Decant the medium and resuspend the cells in 0.5 mL of fresh prewarmed medium. Transfer the cells to the equilibrated plate, dividing them equally between the four center wells. Incubate overnight at 37°C with 5%–10% CO_2.
4. The next day, feed the cells by transferring fresh medium from the neighboring wells to the four wells containing cells. In addition, add fresh medium to eight of the empty wells.

   The cells should be 30%–50% confluent.
5. When the four wells are confluent (at 1–3 d), use a pipette to take up the cells and split them into the eight wells containing medium. In addition, add 50 mL of medium to each of three 150-cm^2 flasks, and 10 mL of medium to a 75-cm^2 flask (which will be used to freeze down aliquots of cells).
6. When the eight wells are almost confluent (at ~1 additional day), pipette up and down to resuspend the cells and transfer them to a 15-mL tube. Transfer 2.5 mL of cells to each of the preequilibrated 150-cm^2 flasks, and 500 µL to the 75-cm^2 flask.
7. To freeze the cells in the 75-cm^2 flask:
   i. Feed the cells every day until they are almost confluent.
   ii. Remove the cells by pipetting up and down and transfer them to a 50-mL tube. Centrifuge for 10 min at 1000 rpm. Meanwhile, count the cells and adjust the density to 2×10^6 cells/mL with a mixture containing 70% RPMI, 20% FCS, and 10% DMSO. Aliquot 1 mL per cryovial.
   iii. Freeze using a cryojar filled to the line with isopropanol, following the manufacturer directions. Alternatively, freeze at −30°C for 3–4 h, then transfer to a −80°C freezer.

   Cells can be left at −80°C for short-term storage. For long term storage, cells must be put in liquid nitrogen.
8. Feed the cells in the 150-cm^2 flasks every day or so (depending on cell density) by doubling the medium. Agitate the cells once a day.
9. Once the cells are nearly confluent, fill the whole flask with medium and stand it up. Let the flask incubate, standing up, for another 2 wk or until most of the cells are dead. Agitate occasionally.
10. To harvest the supernatant, combine the medium from the three flasks in two sterilized 1-L bottles. Add 1/20 volume of 1 M Tris buffer (pH 8) to each bottle and mix well.

    The medium should turn red.
11. Divide the supernatant into 50-mL tubes, and centrifuge for 10 min at 1000 rpm.
12. Transfer the cleared supernatant to 15-mL tubes. Store the aliquots at −30°C.

Cite this protocol as *Cold Spring Harb Protoc*; doi:10.1101/pdb.prot074914

RECIPES

Hybridoma Cell Line Medium

1. Transfer 150 mL of RPMI (Invitrogen 21870-100) from a 500-mL bottle. (Save the remaining 350 mL for Step 3.)
2. Add the following to the 150 mL of RPMI:

   58 mL of fetal calf serum (Gibco 10437-028), heat-inactivated for 30 min at 55°C
   5.8 mL of 100× penicillin/streptomycin (Gibco 15140-122)
   5.8 mL of 0.5 mg/mL insulin stock <R>
   5.8 mL of 100 mM sodium pyruvate (Gibco 11360-070)
   5.8 mL of 200 mM L-glutamine (Gibco 25030-081)
   0.5 mL of 0.1% β-mercaptoethanol

3. Filter to sterilize and add the solution back to the original bottle containing 350 mL of RPMI. Store at 4°C.

Insulin Stock (0.5 mg/mL)

To 20 mL of sterile water, add 10 mg of insulin (Sigma-Aldrich I6634) and 100 µL of 1.0 N HCl. Mix well. Filter through a 0.22-µm filter. Store at 4°C for 4–6 wk.

CHAPTER 2

Purification and Culture of Corticospinal Motor Neurons

Wim Mandemakers[1]

Laboratory for the Research of Neurodegenerative Diseases, VIB Center for the Biology of Disease, KU Leuven Center for Human Genetics, 3000 Leuven, Belgium

Corticospinal motor neurons (CSMNs) residing in cortical layer V of the mammalian brain project their axons to the spinal cord, where they connect with spinal motor neurons (SMNs) located in the ventral horn of the spinal cord. CSMNs and SMNs control voluntary movements, and their importance becomes obvious in situations where this network breaks down (i.e., in amyotrophic lateral sclerosis [ALS] and after spinal cord injury). Here we provide an overview of recent progress in the anatomical, morphological, and genetic characterization of developing CSMNs, as well as their survival requirements. We also describe model systems used to study CSMNs and introduce an immunopanning procedure for the purification and culture of CSMNs. Although these procedures have so far been used to purify only rodent CSMNs, in principle they should work to purify CSMNs from any vertebrate species, as well any type of central nervous system (CNS) or peripheral nervous system (PNS) neuron that can be retrograde labeled.

CSMN ANATOMICAL AND MORPHOLOGICAL CHARACTERIZATION

CSMNs are glutamatergic neurons characterized by their pyramidal shape. They belong to the corticofugal–subcerebral projection class of cortical projection neurons and have secondary collaterals to the striatum, red nucleus, pons, and medulla (Molyneaux et al. 2007). The majority of CSMN neurons in the rat are born between embryonic days 15 and 16 (Bayer et al. 1991) and migrate to layer V by E19–E20 (Miller 1987). During or shortly after CSMNs have reached their final position, they begin to grow axons toward their targets. Initially, not only layer V neurons in the sensorimotor cortex, but also layer V neurons from other cortical areas (i.e., the visual cortex), project their axons to the spinal cord. In contrast to these earlier stages, in adults CSMN neurons are confined to approximately the rostral two-thirds of the cortex, and none are found within the visual cortex (Stanfield et al. 1982; Miller 1987; O'Leary and Terashima 1988). Importantly, this developmental exclusion of CSMN neurons from the visual cortex, as well as from other neocortical areas that do not contain pyramidal tract neurons in the adult, is not the result of cell loss, but rather is due to the selective elimination of the pyramidal tract axons (Stanfield et al. 1982).

CSMN GENERATION AND DEVELOPMENT

During development, CSMNs are generated from progenitors of the neocortical germinal zone located in the dorsal wall of the telencephalon (McConnell 1991; Anderson et al. 2002). CSMN neurons are one of the earliest-generated cortical neurons that migrate away from the germinal zone. Early-born cortical plate cells populate the deepest cortical layers (i.e., layer VI and V), and later-generated

[1]Correspondence: wim.mandemakers@med.kuleuven.be

Cite this introduction as *Cold Spring Harb Protoc;* doi:10.1101/pdb.top070938

neurons migrate past older cells and settle into progressively more superficial positions (Allendoerfer and Shatz 1994).

Transplantation studies suggest that neurons, including CSMNs, acquire a laminar identity, which specifies the layer to which they will migrate, by the time of their terminal mitotic division (Jessell 2000; Shirasaki and Pfaff 2002). Recent studies show that this laminar identity is specified by differential gene expression during the various steps of cortical laminar specification. For instance, expression of the genes *Svet1* and *Cux2* in early progenitor cells marks the cells that will populate the upper-cortical layers (layers II–IV), as opposed to the *Otx1* and *Fezl* expression that marks progenitors that will give rise to lower-cortical layer V and VI neurons (Molyneaux et al. 2007). Further, progressive specification of cortical layers and subpopulations of neurons within specific cortical layers is marked by expression of other genes. For instance, CSMNs are specifically marked by expression of *Diap3*, *Igfbp4*, and *Crim1* within layer V, in contrast to *Ctip2*, *Fezl*, *Clim1*, *S100a10*, and *Pcp4*, which are expressed in all subcerebral projection neurons in layer V, including CSMNs (Arlotta et al. 2005). However, marking a specific population does not necessarily mean that a gene plays an important role in the laminar specification of the cortex. Nevertheless, recent data show that various genes that mark particular cortical neuronal populations do play a key role in the progressive laminar specification of the cortex. For instance, in vivo loss-of-function studies show that *Fezl* is required for the specification of all subcerebral projection neurons, including CSMNs (Chen et al. 2005a,b; Molyneaux et al. 2005).

CSMN FUNCTION AND SURVIVAL IN MODEL SYSTEMS

Several studies have focused on the survival requirements of CSMNs. It has been reported that several growth factors, including brain-derived neurotrophic factor (BDNF), glial cell line–derived growth factor (GDNF), nerve growth factor (NGF), vascular endothelial growth factor (VEGF), leukemia inhibitory factor (LIF), and neurotrophin-3 (NT3), might all promote the survival of CSMNs in vitro and in vivo (Giehl 2001; Lu and Tuszynski 2008). However, most studies have been performed in mixed cultures or in vivo, making it difficult to distinguish direct and indirect effects. Moreover, none of these factors can support long-term CSMN survival. An elegant study by Ozdinler and Macklis (2006) showed that insulin-like growth factor 1 (IGF-1) and BDNF support survival of CSMNs that have been highly purified by fluorescence-activated cell sorting (FACS). Interestingly, both factors appear to have functions independent of their survival promoting activity; IGF-1 specifically enhances the extent and rate of CSMN neurite outgrowth, as opposed to BDNF, which promotes branching and arborization. In spite of this great progress in knowledge of the anatomical, morphological, and genetic characterization of developing CSMNs and their survival requirements, it is clear that more research is needed to investigate the cellular and molecular controls of CSMN survival and axon outgrowth. Such research could lead to a better understanding of diseases involving CSMNs and development of therapies, but progress in the field has been limited by the inability to separate most individual neuronal populations from neighboring neuronal and glial cell types.

Although several groups have succeeded in purifying neuronal populations via immunopanning or FACS, these procedures have some significant drawbacks. Immunopanning is relatively inexpensive, gentle on cells, and can produce very good yields (Barres et al. 1988). However, immunopanning is limited by the requisite availability of an antibody that recognizes a cell surface epitope present exclusively on the cell population of interest, and few if any monoclonal antibodies have been developed that are neuron type–specific. FACS is a more versatile technique, because specific populations can be labeled by genetically encoded or retrograde-transmitted fluorescent tracers, but FACS requires the availability of expensive machinery and often adversely affects the viability of fragile cells such as neurons.

In the accompanying protocols, we describe a novel purification method that combines the versatility of retrograde labeling with the advantages of immunopanning. We inject targeted axonal tracts with cholera toxin B subunit (CTB) adsorbed to fluorescent beads to retrograde-label a neuronal population of interest (see Protocol 1: Retrograde Labeling of Corticospinal Motor Neurons from

 Cite this introduction as *Cold Spring Harb Protoc*; doi:10.1101/pdb.top070938

Early Postnatal Rodents [Mandemakers (a)]). The tissue containing the labeled neuron cell bodies is then dissected and dissociated, and the neurons of interest are specifically purified via immunopanning with an anti-CTB antibody. It is surprising that this method works, but some of the retrograde-transported CTB is trafficked back to the plasma membrane of the neuronal soma. Our method can be used to highly purify any neuronal cell type whose projections can be selectively retrograde-labeled. We have demonstrated the efficacy of this method by purifying to >99% purity two populations of CNS neurons: retinal ganglion cells and CSMNs from early postnatal rat pups (Dugas et al. 2008). Only the purification and culture of CSMNs is described in the companion protocol (see Protocol 2: Immunopanning of Retrograde-Labeled Corticospinal Motor Neurons from Early Postnatal Rodents [Mandemakers (b)]).

REFERENCES

Allendoerfer KL, Shatz CJ. 1994. The subplate, a transient neocortical structure: Its role in the development of connections between thalamus and cortex. *Annu Rev Neurosci* **17:** 185–218.

Anderson SA, Kaznowski CE, Horn C, Rubenstein JL, McConnell SK. 2002. Distinct origins of neocortical projection neurons and interneurons in vivo. *Cereb Cortex* **12:** 702–709.

Arlotta P, Molyneaux BJ, Chen J, Inoue J, Kominami R, Macklis JD. 2005. Neuronal subtype-specific genes that control corticospinal motor neuron development in vivo. *Neuron* **45:** 207–221.

Barres BA, Silverstein BE, Corey DP, Chun LL. 1988. Immunological, morphological, and electrophysiological variation among retinal ganglion cells purified by panning. *Neuron* **1:** 791–803.

Bayer SA, Altman J, Russo RJ, Dai XF, Simmons JA. 1991. Cell migration in the rat embryonic neocortex. *J Comp Neurol* **307:** 499–516.

Chen B, Schaevitz LR, McConnell SK. 2005a. Fezl regulates the differentiation and axon targeting of layer 5 subcortical projection neurons in cerebral cortex. *Proc Natl Acad Sci* **102:** 17184–17189.

Chen JG, Rasin MR, Kwan KY, Sestan N. 2005b. Zfp312 is required for subcortical axonal projections and dendritic morphology of deep-layer pyramidal neurons of the cerebral cortex. *Proc Natl Acad Sci* **102:** 17792–17797.

Dugas JC, Mandemakers W, Rogers M, Ibrahim A, Daneman R, Barres BA. 2008. A novel purification method for CNS projection neurons leads to the identification of brain vascular cells as a source of trophic support for corticospinal motor neurons. *J Neurosci* **28:** 8294–8305.

Giehl KM. 2001. Trophic dependencies of rodent corticospinal neurons. *Rev Neurosci* **12:** 79–94.

Jessell TM. 2000. Neuronal specification in the spinal cord: Inductive signals and transcriptional codes. *Nat Rev Genet* **1:** 20–29.

Lu P, Tuszynski MH. 2008. Growth factors and combinatorial therapies for CNS regeneration. *Exp Neurol* **209:** 313–320.

Mandemakers W. 2014a. Retrograde labeling of corticospinal motor neurons from early postnatal rodents. *Cold Spring Harb Protoc* doi: 10.1101/pdb.prot074922.

Mandemakers W. 2014b. Immunopanning of retrograde-labeled corticospinal motor neurons from early postnatal rodents. *Cold Spring Harb Protoc* doi: 10.1101/pdb.prot074930.

McConnell SK. 1991. The generation of neuronal diversity in the central nervous system. *Annu Rev Neurosci* **14:** 269–300.

Miller MW. 1987. The origin of corticospinal projection neurons in rat. *Exp Brain Res* **67:** 339–351.

Molyneaux BJ, Arlotta P, Hirata T, Hibi M, Macklis JD. 2005. *Fezl* is required for the birth and specification of corticospinal motor neurons. *Neuron* **47:** 817–831.

Molyneaux BJ, Arlotta P, Menezes JR, Macklis JD. 2007. Neuronal subtype specification in the cerebral cortex. *Nat Rev Neurosci* **8:** 427–437.

O'Leary DD, Terashima T. 1988. Cortical axons branch to multiple subcortical targets by interstitial axon budding: Implications for target recognition and "waiting periods". *Neuron* **1:** 901–910.

Ozdinler PH. Macklis JD. 2006. IGF-I specifically enhances axon outgrowth of corticospinal motor neurons. *Nat Neurosci* **9:** 1371–1381.

Shirasaki R. Pfaff SL. 2002. Transcriptional codes and the control of neuronal identity. *Annu Rev Neurosci* **25:** 251–281.

Stanfield BB, O'Leary DD, Fricks C. 1982. Selective collateral elimination in early postnatal development restricts cortical distribution of rat pyramidal tract neurones. *Nature* **298:** 371–373.

Protocol 1

Retrograde Labeling of Corticospinal Motor Neurons from Early Postnatal Rodents

Wim Mandemakers[1]

Laboratory for the Research of Neurodegenerative Diseases, VIB Center for the Biology of Disease, KU Leuven Center for Human Genetics, 3000 Leuven, Belgium

Immunopanning of viable neurons requires availability of an antibody directed against a neuronal surface epitope that is specific for a neuron type of interest (e.g., corticospinal motor neurons [CSMNs]) and that is able to immobilize only neurons of interest from a dissociated brain suspension on a culture dish. However, few, if any, neuron type–specific monoclonal antibodies have been developed. Panning antibodies have not been characterized for CSMNs. To circumvent this issue, we developed a method to retrograde label CSMNs with a cholera toxin B subunit (CTB) tracer that localizes to the plasma membrane only of labeled CSMNs. The main objective of this procedure is to provide CSMNs with a surface marker that can be used to immunopan only the labeled CSMNs by using a CTB-specific antibody.

MATERIALS

It is essential that you consult the appropriate Material Safety Data Sheets and your institution's Environmental Health and Safety Office for proper handling of equipment and hazardous material used in this protocol.

Reagents

Buprenorphine (Buprenex, Reckitt and Colman Pharmaceuticals)

Cholera toxin B subunit (CTB) (List Biological)

Prepare a solution of 1 mg/mL CTB in Dulbecco's phosphate-buffered saline (D-PBS; Invitrogen 14190-094).

Isoflurane (VetEquip)

Lactated Ringer's solution (Fisher Scientific)

Lumafluor Retrobeads IX (Lumafluor)

Rats (Sprague Dawley, P1-P8)

Equipment

Aluminum foil

Forceps (Dumont #7) (Fine Surgical Tools)

GyroMini nutating mixer (Labnet)

Heating pad (Fine Surgical Tools)

Microcentrifuge tubes (0.5 mL)

Needle holder (Fine Surgical Tools)

[1]Correspondence: wim.mandemakers@med.kuleuven.be

Cite this protocol as *Cold Spring Harb Protoc;* doi:10.1101/pdb.prot074922

Scalpel (Fine Surgical Tools)
Scalpel blade (#10) (Fine Surgical Tools)
Stereotactic device (optional; see Step 8)
Surgical glue (Vetbond) (3M; Fisher Scientific)
Syringe (Hamilton 5 µL) with a 32-gauge needle (Fisher Scientific)
Ultracentrifuge (benchtop) (e.g., Beckman Optima TLX with TLA 120.2 rotor)
Vicryl suture (4-0) (VWR)

METHOD

Preparation of CTB/Retrobeads IX

1. Pipette 20 µL of Lumafluor Retrobeads IX into a 0.5-mL microcentrifuge tube and combine with 100 µL of 1 mg/mL CTB.
2. Cover the tube with aluminum foil and incubate on a nutating mixer overnight at 4°C.
3. On the next day, isolate the microspheres in a benchtop ultracentrifuge by spinning at 127,814*g* for 60 min at 4°C.
4. Remove 100 µL of the supernatant and resuspend the pellet in the remaining 20 µL of CTB solution.

   Store at 4°C until use.

Injection of CTB/Retrobeads IX into Rodents

Before beginning surgery, make sure that all procedures have been approved by the relevant animal welfare committees.

5. Anesthetize rats using isoflurane.
6. Using a scalpel and blade, make an incision in the skin at the back of the skull.
7. Using forceps, reflect the muscle tissue and the atlantooccipital membrane.

   This exposes the underlying dura.
8. Lower a 5-µL Hamilton syringe (manually or using a stereotactic device) using a 32-gauge needle in between the foramen magnum and the atlas into the pyramidal decussation at the junction of the medulla and cervical spinal cord.
9. Inject 1 µL of CTB/Retrobeads IX over a time span of 1 min. After the injection, slowly retract the needle.
10. Suture the muscle tissue with Vicryl suture thread 4-0 and close the skin using surgical glue.
11. After surgery, inject pups subcutaneously with lactated Ringer's solution (5 mL/100 g of body weight) and 0.05 mg/kg buprenorphine. Keep pups warm on a heating pad and observe them closely until they are fully alert.
12. After recovery, return the animals to their mother for 2–3 d to allow complete retrograde transport of the injected tracer (see Fig. 1).

   Purify CSMN via immunopanning with an anti-CTB antibody as described in Protocol 2: Immunopanning of Retrograde-Labeled Corticospinal Motor Neurons from Early Postnatal Rodents (Mandemakers).

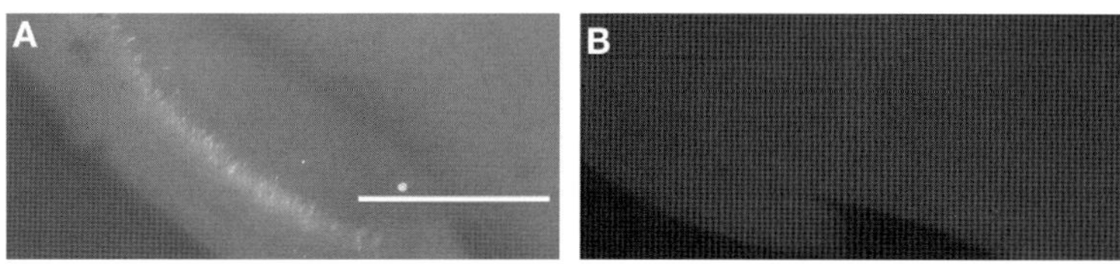

FIGURE 1. (*A*) Successfully labeled dissected cortices can easily be discerned from (*B*) unsuccessful injections that fail to label layer V using an inverted fluorescent microscope.

REFERENCE

Mandemakers W. 2014. Immunopanning of retrograde-labeled corticospinal motor neurons from early postnatal rodents. *Cold Spring Harb Protoc* doi: 10.1101/pdb.prot074930.

Cite this protocol as *Cold Spring Harb Protoc*; doi:10.1101/pdb.prot074922

Protocol 2

Immunopanning of Retrograde-Labeled Corticospinal Motor Neurons from Early Postnatal Rodents

Wim Mandemakers[1]

Laboratory for the Research of Neurodegenerative Diseases, VIB Center for the Biology of Disease, KU Leuven Center for Human Genetics, 3000 Leuven, Belgium

Here we describe a method to purify corticospinal motor neurons (CSMNs). It combines the versatility of retrograde labeling with the advantages of immunopanning. Rat cortices with CSMNs that have been labeled with cholera toxin beta (CTB) are dissected and dissociated, and the CSMNs are specifically purified via immunopanning with an anti-CTB antibody. We show the efficacy of this method in early rat pups by purifying CSMNs to >99% purity. The method can be used to highly purify any neuronal cell type whose projections can be selectively labeled via retrograde tracing.

MATERIALS

It is essential that you consult the appropriate Material Safety Data Sheets and your institution's Environmental Health and Safety Office for proper handling of equipment and hazardous material used in this protocol.

RECIPES: Please see the end of this protocol for recipes indicated by <R>. Additional recipes can be found online at http://cshprotocols.cshlp.org/site/recipes.

Reagents

Antibodies for immunopanning

- Goat anti-mouse IgG, Fcγ fragment-specific (Jackson ImmunoResearch 115-005-164)
- Mouse anti-CTB (mouse IgG1 monoclonal, clone 3D11; Biodesign International)

Anti-oxidants (AO) (1000×) <R>

APV stock (25 mM) <R>

Bovine serum albumen (BSA) stock (4%) <R>

CSMN growth medium, freshly prepared <R>

CSMN survival medium <R> can also be used.

DNase

To prepare a 0.4% stock of DNase in Earle's balanced salt solution (EBSS), add 1 mL of EBSS (Sigma-Aldrich E6267) per 12,500 units of DNase I (Worthington LS002007). Keep on ice. Filter-sterilize, and store in 200-µL aliquots at −20°C.

Dulbecco's modified Eagle medium (DMEM; Invitrogen 11960-044)

Dulbecco's phosphate-buffered saline (D-PBS; Invitrogen 14190-094)

Earle's Balanced Salt Solution (EBSS) (Sigma-Aldrich E6267)

[1]Correspondence: wim.mandemakers@med.kuleuven.be

Cite this protocol as *Cold Spring Harb Protoc*; doi:10.1101/pdb.prot074930

Fetal calf serum (FCS) (Gibco 10437-028)

Prepare 50-mL aliquots of FCS. Heat-inactivate aliquots for 30 min at 55°C, and then store at −20°C.

High-ovomucoid (HI) solution (6×) <R>
Insulin stock (0.5 mg/mL) <R>
Ky stock (0.8 M) <R>
L-cysteine hydrochloride monohydrate (Sigma-Aldrich C7880)
Laminin (mouse) (Cultrex; R&D Systems 3400-010-01)

Thaw mouse laminin (1 mg/mL) at 4°C. Make 10-µL aliquots and store at −30°C.

Lectin from *Griffonia simplicifolia* (BSL-1; Vector Labs L-1100)

Prepare a 2.5 mg/mL BSL-1 solution in D-PBS.

Low-ovomucoid (LO) solution (10×) <R>
NB-sucrose buffer, freshly prepared <R>
NBS buffer, freshly prepared <R>
Neurobasal solution (NB) (Invitrogen 21103-049)
Papain (Worthington LS03126)
Poly-D-lysine (PDL) (100×)

Add 5 mL of H_2O to a 5-mg bottle of poly-D-lysine (PDL; Sigma-Aldrich P6407). Filter through a 2-µm filter. Make 100-µL aliquots and store at −20°C. Dilute stock to 1× in sterile H_2O before use.

Rat pups from Protocol 1: Retrograde Labeling of Corticospinal Motor Neurons from Early Postnatal Rodents (Mandemakers)
Tris-HCl (50 mM, pH 9.5) (sterile)

Dissolve 12.1 g of Trizma base in 200 mL of dH_2O. Adjust pH to 9.5 with HCl.

Trypsin stock (30,000 U/mL)

Dissolve trypsin (Sigma-Aldrich T9935) at 30,000 U/mL in EBSS. Filter through a 0.22-µm filter. Make 200-µL aliquots and store at −80°C. Note that trypsin may lose activity, even when stored at −80°C.

Equipment

Centrifuge (tabletop, with 15- and 50-mL conical tube adaptors)
Dissection microscope
Ethanol-washed coverslips <R>
Forceps (sterile)
Heat block base (set to 34°C in sterile hood)
Hemocytometer
Nitex mesh filters

Cut the mesh into 3 inch squares, wrap in small packets of foil and autoclave.

Petri dishes (6-, 10-, and 15-cm) (Falcon or Nunc)

Make a small hole in the center of a 6-cm dish lid that can accommodate a 0.22-µm filter. The hole can be created by melting an opening into the center of the lid using flamed forceps.

Pipette (glass, fire-polished)
Plastic bags (resealable) (to hold single 15-cm Petri dishes)
Scalpel (Fine Surgical Tools)
Scalpel blade (#10, curved-edge) (Fine Surgical Tools)
Source of 5% CO_2/95% O_2 with a line leading to a 34°C heat block in a sterile hood
Syringe filters (0.22-µm)
Tissue culture plates (plastic, 24-well) (Falcon or Nunc)

Make a small hole in the center of a 24-well plate lid that can accommodate a 0.22-µm filter. The hole can be created by melting an opening into the center of the lid using flamed forceps.

Tubes (50-mL conical)
Tubes (30-mL universal)
Water baths (34°C, 37°C)

Cite this protocol as *Cold Spring Harb Protoc*; doi:10.1101/pdb.prot074930

METHOD

An overview of retrograde labeling followed by immunopanning for CSMN purification is provided in Figure 1.

Preparation

Perform Steps 1–6 on the first day and the remaining steps on the following day.

1. Prepare the panning dishes as follows.
   i. To prepare the BSL-1 panning dishes, coat four 15-cm Petri dishes with 20 mL of NBS buffer and 20 µL of BSL-1 per dish.
   ii. To prepare a CTB panning dish, coat a single 10-cm Petri dish with 40 µL of goat anti-mouse IgG Fcγ fragment-specific antibody and 10 mL of 50 mM Tris-HCl (pH 9.5).
2. Incubate the dishes overnight on a flat surface at 4°C.
   Plates are initially hydrophobic, but after coating overnight, plates become visibly hydrophilic. If plates are needed immediately, a quick but not ideal way of preparing plates is to coat with the secondary antibodies for 2 h at 37°C. For best results, however, coat the plates with secondary antibodies overnight.
3. Prepare the coverslips as follows.
   i. Rinse the ethanol-washed coverslips with sterile H_2O in a 15-cm Petri dish. Use sterilized forceps to transfer each coverslip into a well of a 24-well plate.
   ii. Add 500 µL of 1× poly-D-lysine in sterile H_2O to each coverslip. Incubate at room temperature for at least 30 min.
   iii. Rinse the coverslips three times with sterile H_2O.
   iv. Dilute laminin to 2 µg/mL (1×) in 50:50 DMEM:NB. Add 500 µL of 1× laminin solution to each coverslip.
4. Incubate the coverslips overnight at 37°C, 5% CO_2.
5. Label tubes for the dissection and dissociation buffers as follows.

   Dissection buffer

   i. Prepare two 50-mL tubes each containing 50 mL of NB-sucrose and label as "Dissection Buffer."
      Dissection buffer will be prepared in Step 14.

   Dissociation buffers

   ii. Prepare one 50-mL tube containing 40 mL of NB-sucrose and label as "Papain Solution."
      Papain solution will be prepared in Step 21.
   iii. Prepare four 50-mL tubes each containing 27 mL of NB-sucrose and label as "LO Solution."
      LO solution will be prepared in Step 26.
   iv. Prepare two 50-mL tubes each containing 24 mL of NB-sucrose and label as "HI Solution."
      HI solution will be prepared in Step 26.
6. Equilibrate the buffers from Step 5 by placing the tubes in a rack overnight at 37°C with the lids slightly opened.
   The buffers will be equilibrated and ready to use on the day of the purification.
7. Equilibrate the NB for the trypsin reaction by placing one universal tube containing 10 mL of NB at 37°C with the lid slightly opened.
   This will be used to remove the cells from the final panning dish.

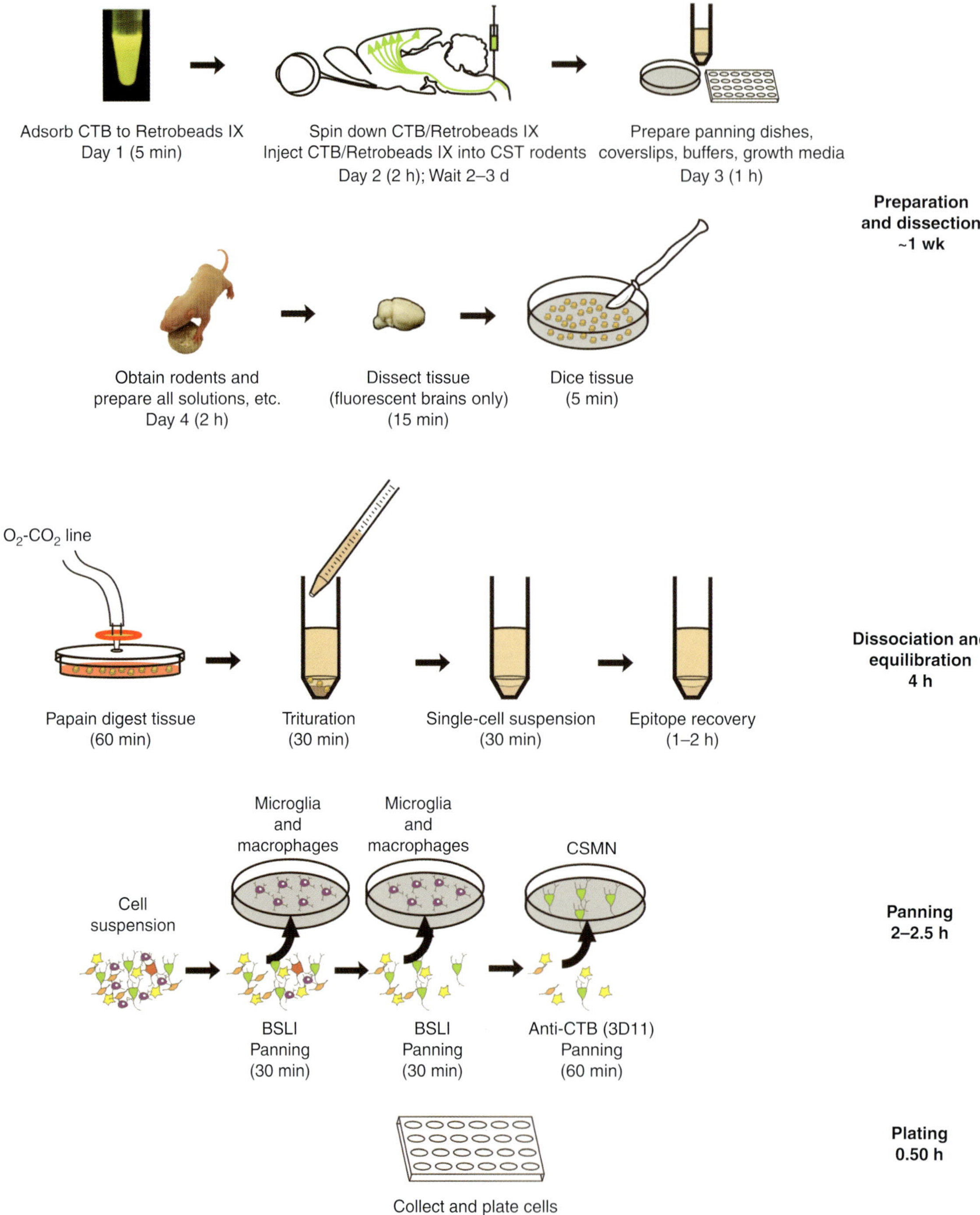

FIGURE 1. Purification of CSMN from early postnatal rat pups by retrograde labeling (as described in Protocol 1: Retrograde Labeling of Corticospinal Motor Neurons from Early Postnatal Rodents [Mandemakers]) and subsequent immunopanning (as described here).

8. Prepare the NBS/0.2% BSA solution by combining 38 mL of NBS buffer and 2 mL of 4% BSA in a 50-mL tube.
9. Prepare the panning buffer by combining 27 mL of NBS buffer, 3 mL of NBS/0.2% BSA, and 300 µL of insulin in a 50-mL tube.
10. Coat the CTB panning dishes with the primary antibody as follows.

 Cite this protocol as *Cold Spring Harb Protoc*; doi:10.1101/pdb.prot074930

i. Wash the CTB panning dish prepared in Step 1.ii three times with NBS buffer.

ii. Add the mouse anti-CTB antibody (140 µg for P3 pups; 210 µg for P7 pups, diluted in 10 mL of NBS/0.2% BSA) to the dish and set aside at room temperature.

iii. Swirl the plate to coat. Incubate at room temperature for at least 2 h.

iv. Just before use, rinse the plate three times with NBS buffer.

11. Leave the BSL-1 dishes prepared in Step 1.i as they are. Wash them three times with NBS just before use (Step 49).
12. Weigh 3.2 mg of L-cysteine, and store until use.
13. Prepare FCS buffer by combining 9.5 mL of NB and 4 mL of heat-inactivated FCS. Filter through a rinsed 0.22-µm filter.

Dissection

14. Take the tubes containing NB-sucrose for dissection buffer (Step 5.i) out of the incubator. Prepare dissection buffer by adding 100 µL of APV stock, 50 µL of Ky stock, and 50 µL of AO to each tube. Close the lids tightly to keep the buffer equilibrated.
15. Prepare one 6-cm dish containing 7 mL of the equilibrated dissection buffer.
16. Prepare a 24-well dish by filling twice the number of wells as the number of cortices to be dissected with 1 mL of dissection buffer per well.

    Hemispheres will be collected one per well; the possibility exists that only one hemisphere is labeled.
17. Leave the 24-well dish in the hood under a stream of 95% O_2/5% CO_2.
18. Dissect the cortices from the rat pups and collect them in the 6-cm dish containing 7 mL of equilibrated dissection buffer.

    For a protocol that includes detailed brain dissection procedures, see Chapter 5, Protocol 1: Purification of Rat and Mouse Astrocytes by Immunopanning (Foo).
19. Clean out the cortices by removing the midbrain, hippocampus, olfactory bulbs, pia, etc. Collect one hemisphere per well in the 24-well dish.
20. Check the cortices under a fluorescent microscope for the presence of tracer. Only use cortices that have tracer present.

    See Figure 1 in Protocol 1: Retrograde Labeling of Corticospinal Motor Neurons from Early Postnatal Rodents (Mandemakers).

Dissociation

Here, NaCl is replaced by sucrose in the enzyme and inhibitor solutions. The rationale for this modification is discussed in terms of acute neurotoxic effect of passive chloride entry and subsequent cell swelling and lysis (Aghajanian and Rasmussen 1989).

From this point onward the protocol is described for eight labeled cortical hemispheres. When using a different number of hemispheres, volumes should be adjusted accordingly; incubation times and temperatures can remain the same.

21. Prepare the papain solution.

i. Add 800 units of papain to the NB-sucrose buffer equilibrated for the enzyme solution in Step 5.ii.

ii. Add 3.2 mg of L-cysteine and incubate in a water bath at 34°C for at least 10 min, tightly capped, to dissolve and activate papain.

The temperature must be at least 30°C for the enzyme to completely dissolve.

iii. Filter the dissolved papain solution through a 0.22-µm syringe filter to remove yeast. Add 80 µL of APV (final concentration, 0.05 mM), 40 µL of AO, 40 µL of Ky (final concentration, 0.8 mM), and 400 µL of DNase I.

10 mL of this solution can be used for up to two rat cortices.

22. Prepare four 6-cm dishes, each with a drop of dissection buffer.
23. Place the cleaned and labeled cortices in the dishes, with up to two rat cortices per drop of dissection buffer. Chop the tissue into ~1-mm^2 slices with a curved-edge scalpel.
24. Add 10 mL of the papain solution to each dish containing cortices and place the dishes on a wetted towel on a heat block at 34°C.
25. Incubate the tissue slices in the papain solution for 60 min at 34°C under a stream of 95% O_2/5% CO_2, gently stirring every 10 to 15 min. (Take care not to bubble the solution!)
26. Fifteen minutes before the end of the papain reaction, take the remaining tubes of NB-sucrose equilibrated for dissociation (Step 5.iii–5.iv) out of the incubator. For each tube of LO solution, add 3 mL of LO frozen stock. For each tube of HI solution, add 6 mL of HI frozen stock. Close the lids and let the solutions adjust to room temperature in the hood.
27. Just before use, add the following to each tube of LO solution: 60 µL of APV, 30 µL of AO, 30 µL of Ky, and 600 µL of DNase I. Add the following to each tube of HI solution: 60 µL of APV, 30 µL of AO, and 30 µL of Ky.
28. After the 60-min incubation (Step 25), pour the tissue from each 6-cm dish into a separate universal tube, allow the tissue pieces to settle, and carefully aspirate as much papain solution as possible without disturbing the tissue pieces.
29. Gently wash the tissue three times with 1 mL of LO solution per wash, letting each wash stand for 1 min to allow the DNase I to work and the tissue pieces to settle.
30. In the meantime, prepare four 50-mL tubes by adding 3 mL of LO solution to each tube.

    These 50-mL tubes will be used to collect the cell suspension from each universal tube.
31. Add 3 mL of LO solution to each universal tube, and triturate the tissue very slowly (~10 times up and down) using a 25-mL pipette. Allow the nondissociated pieces to settle, draw off the dissociated cells in the supernatant and place them into a 50-mL collection tube containing LO solution.
32. Add fresh LO solution to each tissue sample and repeat until each tube of LO solution is finished. Perform the final two triturations with a fire-polished glass pipette.

    Each of the four 50-mL tubes should now contain ~30 mL of cell suspension.
33. Discard any remaining nondissociated material.
34. Set aside a small sample from each 50-mL tube (5 µL diluted in 1:10 in NB-sucrose to final volume of 50 µL) to count cell number and determine viability.
35. Remove any nondissociated tissue chunks that have settled at the bottom of the tube.
36. For each cell suspension, generate a step gradient by layering the dissociated cells on top of 15 mL of HI solution.

    This can be performed best by drawing up the HI solution in a 10-mL pipette and gently layering it underneath the cell suspension.
37. Centrifuge the cells for 10 min at 1000 rpm at room temperature.
38. Resuspend each pellet in 10 mL of equilibrated dissection buffer.
39. Layer 1 mL of 4% BSA underneath the cell suspension and centrifuge again for 10 min at 1000 rpm at room temperature.

Epitope Recovery and Buffer Transition

40. During centrifugation, add 200 µL of DNase I stock solution, 60 µL of APV (final concentration, 0.05 mM), 30 µL of AO, and 30 µL of Ky (final concentration, 0.8 mM) to the panning buffer prepared in Step 9.

 Cite this protocol as *Cold Spring Harb Protoc*; doi:10.1101/pdb.prot074930

41. Prepare five tubes containing the following buffer solutions.
    i. Tube 1: 0.5 mL dissection buffer + 0.5 mL panning buffer (1-mL total volume)
    ii. Tube 2: 0.17 mL dissection buffer + 0.83 panning buffer (1-mL total volume)
    iii. Tube 3: 0.33 mL dissection buffer + 1.67 panning buffer (2-mL total volume)
    iv. Tube 4: 3 mL panning buffer
    v. Tube 5: 9 mL panning buffer
42. Place the tubes in a 37°C incubator with the lids of Tubes 1–3 slightly open and the lids of Tubes 4 and 5 closed. Check the color of the buffers in Tubes 1–3. If yellow, close the lids.
43. After centrifugation (Step 39), resuspend each pellet in 0.5 mL of equilibrated dissection buffer.
44. Combine two cell suspensions and transfer to a single universal tube.
    There will now be two universal tubes containing 1 mL of cell suspension each.
45. Place the cell suspensions at 37°C with the tube lids slightly opened.
    The cells will remain in the incubator for 60 min to allow epitope recovery. During this period, the buffer will be changed to buffer with NaCl and reduced carbonate concentration (panning buffer).
46. Incubate the cells for a total of 60 min, changing the buffer gradually as follows.
    i. After 10 min, divide the contents of Tube 1 equally between each cell suspension (0.5 mL per tube).
    ii. After 20 min, divide the contents of Tube 2 equally between each cell suspension (0.5 mL per tube).
    iii. After 30 min, divide the contents of Tube 3 equally between each cell suspension (1 mL per tube).
    iv. After 40 min, divide the contents of Tube 4 equally between each cell suspension (1.5 mL per tube).
    v. After 50 min, divide the contents of Tube 5 equally between each cell suspension (4.5 mL per tube).
       The final volume should be ~9 mL per universal tube, and the $NaHCO_3$ concentration will be reduced to 2.9 mM.
    vi. After 60 min, remove the cells from the incubator and place them in a tissue culture hood with the lids closed for 15 min to adjust to room temperature.
47. Pour the cell suspension through a Nitex filter to remove any aggregates that might have formed after epitope recovery.
48. Rinse the tube and the Nitex filter with panning buffer, and proceed with panning.

Panning

49. Wash the first two BSL-1 panning dishes three times each with NBS buffer.
50. Pour the cell suspensions from each universal tube onto separate BSL-1 dishes.
51. Rinse the tubes with NBS buffer.
    At this stage, the total volume is ~12 mL per universal tube and the concentration of $NaHCO_3$ is 2.16 mM.
52. Place each panning dish in a plastic bag. Flush each bag with 95% O_2/5% CO_2 to keep the pH balanced during panning at room temperature. Close the bags and place the dishes on a flat surface.
53. Pan for 30 min at room temperature.
54. Wash the last two BSL-1 panning dishes three times with NBS buffer.

55. Remove the first two BSL-1 plates from the bags, shake the dishes to dislodge any nonattached cells, and pour the cells from each plate onto the second set of BSL-1 panning dishes.
56. Place each BSL-1 panning dish in a plastic bag. Flush each bag with 95% O_2/5% CO_2. Close the bags and place the dishes on a flat surface.
57. Pan for 30 min at room temperature.
58. Wash the CTB panning dish three times with NBS buffer.
59. Remove the second set of BSL-1 plates from the bags, shake the dishes to dislodge any nonattached cells, and combine the cells from both BSL-1 dishes onto the single CTB panning dish (volume ~24 mL).
60. Place the panning dish in a plastic bag. Flush the bag with 95% O_2/5% CO_2. Close the bag and place the dish on a flat surface.
61. Pan for 60 min at room temperature. Perform Steps 62–64 during panning.

Trypsinization and Plating

62. During Step 61, aspirate the laminin and add CSMN growth medium to the wells. Place the plate in the incubator to equilibrate.
63. Allow the FCS buffer prepared in Step 13 to warm to at least room temperature.
64. During the final 5 min of CTB panning (Step 61), prepare the trypsin solution by adding 100 µL of 2.5% trypsin to 4 mL of pre-equilibrated NB.
65. Remove the CTB panning dish from the plastic bag.
66. Gently wash the plate approximately five times with NBS buffer to remove all nonadherent cells.
67. Visually examine the plate under a bright field dissection microscope to confirm that all nonadherent cells have been removed.
68. Add 6 mL of preequilibrated NB to the dish; quickly swirl and discard.
69. Add 4 mL of trypsin solution to the dish. Incubate for 5 min at 37°C.
70. Add 1 mL of FCS buffer to a 50-mL tube labeled "Cells" for transfer of the released cells.
71. Squirt the trypsin solution once from the edge to the center of the dish. Determine, by visual analysis through a bright-field dissection microscope, whether most of cells have been dislodged from the area of interest. If not, return the plate to the incubator to extend the trypsinization time.
72. When trypsinization is satisfactory, add 2 mL of FCS buffer to the cells. Using a P1000 pipette, squirt the cells off the dish, pipetting toward the center of the dish while rotating the dish around once. (It helps one to make a mark on plate for the starting point.) Next, squirt once around the dish, pipetting along the edge.
73. Determine, by visual analysis through a bright-field dissection microscope, whether most of the cells have been dislodged from the plate. Transfer the cells to the 50-mL tube containing 1 mL of FCS buffer.
74. Add another 5 mL of FCS buffer to the plate. Pipette up and down as described in Step 72.
75. Determine, by visual analysis through a bright-field dissection microscope, whether most of the cells have been dislodged from the plate, paying close attention to the edges of the plate. If all the cells have not been released, continue pipetting.
76. When all of the cells are dislodged, add these cells to the previously collected cells in the 50-mL tube. Rinse the plate with the remaining FCS buffer and add this solution to the previously collected cells.

 Cite this protocol as *Cold Spring Harb Protoc*; doi:10.1101/pdb.prot074930

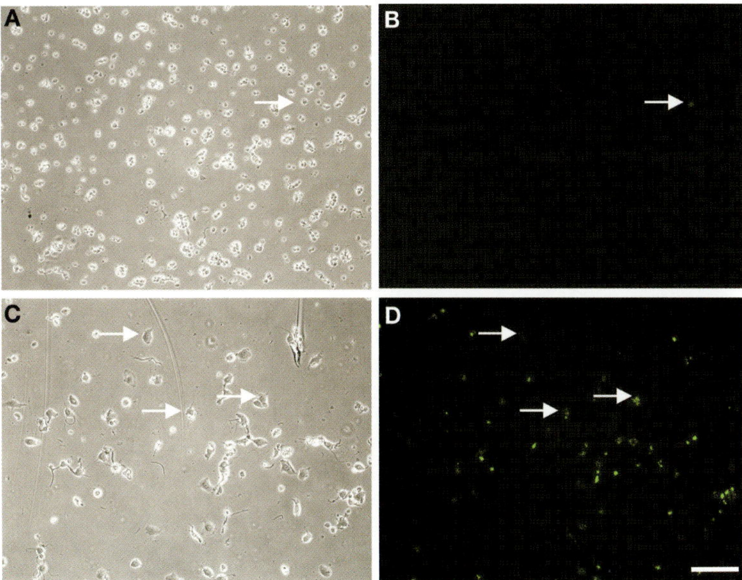

FIGURE 2. Cells before and after CTB panning cultured for 1 d in growth medium. (*A*,*B*) Labeled cells can barely be detected in the cell population before CTB panning. (*C*,*D*) Most cells are labeled with Retrobeads IX in CTB immunopanned cells. Scale bar, 100 µm.

77. Gently mix the cells and remove 50 µL for counting using a hemocytometer.
78. Centrifuge the cells at 1000 rpm for 11 min at room temperature. In the meantime, count the cells.
79. Resuspend the cell pellet in preequilibrated CSMN growth or survival medium and plate the cells at the desired density and culture conditions (Fig. 2).

DISCUSSION

Panning is trivial, involving only three steps: enzymatic preparation of a cell suspension, passing this suspension over a series of antibody-coated dishes, and removing the purified cells from the final dish. In general, the following general guidelines should be followed: Attempt to minimize the length of the entire procedure, and avoid cooling neurons lower than room temperature even for only a few minutes. Trypsin may lose activity even at −80°C—both the aliquots and the stock. This could result in extended incubation times at Steps 69–79 and consequently in reduced survival of the purified neurons. If this occurs, replace the trypsin with a fresh stock. An extended discussion on the technique described here can be found in Dugas et al. (2008).

RECIPES

Anti-Oxidants (AO) (1000×)

1. Add 1.68 mg of gluthatione (Sigma-Aldrich G6013) and 4.2 mg of catalase (Sigma-Aldrich C40) to a stock vial of SOD (Sigma-Aldrich S2515), which contains 4.2 mg.
2. Prepare 0.1% BSA in NBS buffer <R>.
3. Place the DL-α-tocopherol (vitamin E; Sigma-Aldrich T3251) and DL-α-tocopherol acetate (Sigma-Aldrich T3001) stocks in a 37°C water bath to dissolve. Once dis-

solved, weigh 100 mg of each in a separate sterile cryovial. Adjust both stocks to 1 mL with 100% ethanol. Add 16.8 µL of each ethanol stock to the SOD vial and mix.

4. Adjust the final volume in the vial to 1.68 mL with 0.1% BSA in NBS.
5. Make aliquots of 50 µL and store at −30°C.

APV Stock (25 mM)

Dissolve 25 mg of DL-2-amino-5-phosphonopentanoic acid (APV) (Sigma-Aldrich A5282) per 5.073 mL of H_2O. Filter to sterilize. Make aliquots of 100 µL and store at −30°C.

BDNF Stock (50 µg/mL)

1. Prepare a master BDNF stock (1 mg/mL) by resuspending 1 mg of human brain-derived neurotrophic factor in powder form (BDNF; Peprotech 450-02) in 1 mL of cold, sterile 0.2% BSA (Sigma-Aldrich A-4161) that was prepared in D-PBS (Gibco 14287). Make 200-µL aliquots of the master stock, flash-freeze in liquid nitrogen, and store at −80°C.
2. To make a working BDNF stock, thaw a 200-µL aliquot of master stock on ice. At the same time, chill 3.8 mL of sterile 0.2% BSA solution on ice. Once chilled, add the master BDNF stock to the 0.2% BSA solution and mix well, but gently, to avoid foaming. Make 20-µL working aliquots and flash-freeze in liquid nitrogen. Store at −80°C.

BSA Stock (4%)

Dissolve 2 g of BSA (Sigma-Aldrich A4161) in 50 mL of D-PBS at 37°C. Adjust pH to 7.4 with ~200 µL of 1 N NaOH. Filter sequentially through 0.45-µm and 0.22-µm filters. Prepare 1.0-mL aliquots and store at −20°C.

CSMN Growth Medium

Reagent	Amount to add (for 20 mL)
Neurobasal medium (Gibco 21103-049)	9.1 mL
DMEM (Gibco 11960-044)	9.1 mL
Penicillin/streptomycin (100×) (Gibco 15140-122)	200 µL
L-Glutamine (200 mM) (Gibco 25030-081)	200 µL
Glucose (2 M)	350 µL
BSA stock (4%) <R>	17 µL
SATO supplement, NB-based (100×) <R>	200 µL
Sodium pyruvate (100 mM) (Gibco 11360-070)	200 µL
Insulin stock (0.5 mg/mL) <R>	200 µL
NAC stock (5 mg/mL) <R>	20 µL
B-27 supplement	400 µL

Filter through a rinsed 0.22-µm filter to sterilize and store at 4°C. Add growth factors, as desired, just before use. (More details about growth factors can be found in Dugas et al. [2008].) Some lots of B-27 have been toxic; an alternative is NS21 from R&D Systems (AR008).

 Cite this protocol as *Cold Spring Harb Protoc*; doi:10.1101/pdb.prot074930

CSMN Survival Medium

Reagent	Amount to add (for 20 mL)
Neurobasal medium (Gibco 21103-049)	9.5 mL
DMEM (Gibco 11960-044)	9.5 mL
Penicillin/streptomycin (100×) (Gibco 15140-122)	200 µL
L-Glutamine (200 mM) (Gibco 25030-081)	200 µL
Glucose (2 M)	350 µL
BSA stock (4%) <R>	17 µL
SATO supplement, NB-based (100×) <R>	200 µL

Filter through a rinsed 0.22-µm filter to sterilize and store at 4°C. As desired, add the following compounds or growth factors just before use:

B-27 supplement
Anti-oxidants (AO) (1000×) <R>
BDNF (50 µg/mL) <R>

Ethanol-Washed Glass Coverslips

Extensively wash 12-mm glass coverslips (Carolina Biological Supply 633029) in 70% ethanol. Perform the washes on a platform shaker in a beaker, with enough motion to lightly agitate the coverslips but not break too many. Wash the coverslips for about 1 mo, exchanging the ethanol approximately every day. (It is fine to skip some exchanges.) Store the washed coverslips in 70% ethanol until use.

High-Ovomucoid (HI) Solution (6×)

To 20 mL of Dulbecco's phosphate buffered saline (D-PBS; Invitrogen 14190-094), add 1200 mg of BSA (Sigma-Aldrich A8806). Mix well. Add 1200 mg of trypsin inhibitor (Boerhinger Mannheim 109878) and mix to dissolve. (The solution may be left at 4°C overnight to allow reagents to dissolve completely.) Adjust the pH to 7.4 by adding at least 600 µL of 1 N NaOH. Filter through a 0.22-µm Millipore Millex-GP filter or a Millipore SteriFlip. Make 1.0-mL aliquots and store at −20°C.

Insulin Stock (0.5 mg/mL)

To 20 mL of sterile water, add 10 mg of insulin (Sigma-Aldrich I6634) and 100 µL of 1.0 N HCl. Mix well. Filter through a 0.22-µm filter. Store at 4°C for 4–6 wk.

Ky Stock (0.8 M)

Dissolve 1.5136 g kynurenic acid (Ky) (Sigma-Aldrich K3375) per 10 mL of NaOH. Filter to sterilize. Make aliquots of 100 µL and store at −30°C.

Low-Ovomucoid (LO) Solution (10×)

To 40 mL of Dulbecco's phosphate buffered saline (D-PBS; Invitrogen 14190-094), add 600 mg of BSA (Sigma-Aldrich A8806). Mix well. Add 600 mg of trypsin inhibitor (Boerhinger Mannheim 109878) and mix to dissolve. Adjust the pH to 7.4 by adding ~400 µL of 1 N NaOH. When completely dissolved, filter through a 0.22-µm Millipore Millex-GP filter or a Millipore SteriFlip. Make 1.0-mL aliquots, and store at −20°C.

NB-Sucrose Buffer

Reagent	Quantity	Final concentration
Sucrose	26.1 g	152.6 mM
$CaCl_2$	132.3 mg	1.8 mM
$MgCl_2$	81.32 mg	0.8 mM
KCl	201 mg	5.4 mM
Glucose	2.25 g	25 mM
NaH_2PO_4 (see Step 4)	54 mg	0.9 mM
$NaHCO_3$	1.1 g	26 mM
HEPES (1 M)	5 mL	10 mM
Pyruvate	1.15 mL	0.23 mM
Phenol red	0.5 mL	0.0011%

1. In a 500-mL cylinder, add all dry components except for NaH_2PO_4 and $NaHCO_3$.

   This is important so that the carbonate and phosphate do not go out of solution.

2. Add milliQ H_2O to 300 mL and dissolve the salts/sucrose.
3. Add HEPES, phenol red and pyruvate.
4. In a 50-mL tube, dissolve 108 mg of NaH_2PO_4 in 50 mL of milliQ H_2O. Add 25 mL of NaH_2PO_4 to the cylinder containing NB-sucrose.

   Prepare NB-sucrose buffer in parallel with NBS buffer <R>, and use the other 25 mL for the NBS buffer.

5. In another 50-mL tube, dissolve 1.1 g of $NaHCO_3$ in 50 mL of milliQ H_2O. Add 50 mL of $NaHCO_3$ to the cylinder containing NB-sucrose.
6. Bring the NB-sucrose solution to 500 mL with ddH_2O and adjust the pH if necessary.
7. Measure the osmolarity of the NBS and NB-sucrose solutions by testing 20 µL of each on an osmometer.

   The two solutions (NBS and NB-sucrose) should be within 10 mOsm of each other (~260 mOsm).

8. After adding sucrose to the NBS buffer as needed, remeasure the osmolarity of both solutions.
9. Filter to sterilize and store at 4°C.

NBS Buffer

Reagent	Quantity	Final concentration
NaCl	2.99 g	102.3 mM
$CaCl_2$	132.3 mg	1.8 mM
$MgCl_2$	81.32 mg	0.8 mM
KCl	201 mg	5.4 mM
Glucose	2.25 g	25 mM
NaH_2PO_4 (see Step 4)	54 mg	0.9 mM
HEPES (1 M)	5 mL	10 mM
Pyruvate (100 mM)	1.15 mL	0.23 mM
Phenol red	0.5 mL	0.0011%

1. In a 500-mL cylinder, add all dry components except for NaH_2PO_4.

   This is important so that the phosphate does not go out of solution.

2. Add milliQ H_2O to 300 mL and dissolve the salts.

Cite this protocol as *Cold Spring Harb Protoc*; doi:10.1101/pdb.prot074930

3. Add HEPES, phenol red, and pyruvate.
4. In a 50-mL tube, dissolve 108 mg of NaH_2PO_4 in 50 mL of milliQ H_2O. Add 25 mL of NaH_2PO_4 to the cylinder containing NBS.

   Prepare NBS buffer in parallel with NB-sucrose buffer <R>, and use the other 25 mL for the NB-sucrose buffer.

5. Bring the NBS solution to 500 mL with ddH_2O and adjust the pH if necessary.
6. Measure the osmolarity of the NBS and NB-sucrose solutions by testing 20 µL of each on an osmometer. As needed, bring up the osmolarity of the NBS solution by adding sucrose.

   The NBS and NB-sucrose solutions should be within 10 mOsm of each other (~260 mOsm). Typically, the osmolarity of the NBS solution is initially a bit lower than the NB-sucrose. Sucrose is 342.3 g/mol, so if the osmolarity needs to be brought up by 25 mOsm, add 4.28 g of sucrose to 500 mL of NBS (342.3 g/mol × 0.025 mol/L × 0.5 L = 4.28 g).

7. After adding sucrose to the NBS buffer, remeasure the osmolarity of both solutions.
8. Filter to sterilize and store at 4°C.

NAC Stock (5 mg/mL)

To prepare, dissolve 50 mg of *N*-acetyl-L-cysteine (NAC) powder (Sigma-Aldrich A8199) in 10 mL of Neurobasal Medium (Gibco/Life Technologies 21103). (The solution will be yellowish.) Filter through a 0.22-µm filter. Prepare 20- and 80-µL aliquots and store them at 4°C.

SATO Supplement, NB-Based (100×)

1. Prepare the following stock solutions (these should be made fresh; do not reuse).
   - Combine 2.5 mg of progesterone (Sigma-Aldrich P8783) and 100 µL of ethanol to make a progesterone stock solution.
   - Combine 4.0 mg of sodium selenite (Sigma-Aldrich S5261), 10 µL of 1 N NaOH, and 10 mL of Neurobasal (NB, Gibco 21103-049) to make a sodium selenite stock solution.
2. Add the following to 80 mL of Neurobasal medium:

Reagent	Quantity	Final concentration in medium (1×)
BSA (Sigma-Aldrich A4161)	800 mg	100 µg/mL
Transferrin (Sigma-Aldrich T1147)	800 mg	100 µg/mL
Putrescine dihydrochloride (Sigma-Aldrich P5780)	128 mg	16 µg /mL
Progesterone stock solution	20 µL	60 ng/mL (0.2 µM)
Sodium selenite stock solution	800 µL	40 ng/mL

3. Mix well, and filter-sterilize through a prerinsed 0.22-µm filter. Make 200-µL or 800-µL aliquots, and store at −20°C.

ACKNOWLEDGMENTS

This work was supported by National Eye Institute Regeneration Grant RO1 EY011310 and the Adelson Medical Research Foundation. W.M. was supported by National Multiple Sclerosis Society Postdoctoral Fellowship FG 1434-A-1 and Netherlands Organization for Scientific Research TALENT Stipend S 93-380.

REFERENCES

Aghajanian GK, Rasmussen K. 1989. Intracellular studies in the facial nucleus illustrating a simple new method for obtaining viable motoneurons in adult rat brain slices. *Synapse* **3:** 331–338.

Dugas JC, et al. 2008. A novel purification method for CNS projection neurons leads to the identification of brain vascular cells as a source of trophic support for corticospinal motor neurons. *J Neurosci* **28:** 8294–8305.

Foo LC. 2013. Purification of rat and mouse astrocytes by immunopanning. *Cold Spring Harb Protoc* doi: 10.1101/pdb.prot074211.

Mandemakers W. 2014. Retrograde labeling of corticospinal motor neurons from early postnatal rodents. *Cold Spring Harb Protoc* doi: 10.1101/pdb.prot074922.

Cite this protocol as *Cold Spring Harb Protoc*; doi:10.1101/pdb.prot074930

CHAPTER 3

Purification and Culture of Spinal Motor Neurons

David J. Graber[1] and Brent T. Harris[2,3]

[1]*Dartmouth Medical School, Department of Pathology, Lebanon, New Hampshire 03756;* [2]*Georgetown University Medical Center, Departments of Neurology and Pathology, Washington, D.C. 20057*

Motor neurons are responsible for voluntary movement. Lower motor neurons are characterized by large soma, the potential to form very long axons, and wide-ranging dendritic arborization. They receive direction from various neuronal cell types and induce movement of skeletal muscle fibers through acetylcholine release at the neuromuscular junction. Each lower motor neuron can communicate with 10 to several hundred muscle fibers at firing rates modulated by the balance of ongoing neurotransmitter signaling. Disease and trauma that affect lower motor neurons can cause paralysis and, in some cases, death. Studies using primary cultures of these cells have ongoing potential to facilitate a deeper understanding of their biology and function.

MOTOR NEURON CHARACTERISTICS

Motor neurons arise from the ventral portion of the neural tube and are among the first cells born in the spinal cord (Altman and Bayer 1984). A large proportion of these cells undergo apoptosis during postnatal development (Yamamoto and Henderson 1999). Lower motor neurons are located in the ventral horn (anterior horn in human), throughout the spinal cord, and in brain-stem nuclei. They induce voluntary movement via acetylcholine release at the neuromuscular junction on skeletal muscle fibers. Each muscle fiber is supplied by a single axon terminal, and a single motor neuron can form synapses with as few as ten or as many as several hundred muscle fibers. Direction for voluntary movement is received from neurons in supraspinal regions (predominantly the primary motor cortex), interneurons, sensory neurons, and even other lower motor neurons. The sum of ongoing inhibitory (i.e., GABA) and excitatory (i.e., glutamate) signals determines the motor neuron's firing rate and resultant muscle-fiber contraction (Carp and Wolpaw 2001).

The soma of motor neurons is typically two to three times larger than that of most other neurons. In addition to their large soma size, lower motor neurons can have an axon that extends very long distances. A motor axon arising from the lumbar spinal cord with its target muscle in the foot can reach a length of up to 1 m in humans. Paradoxically, the majority of the cell surface area is in its uniquely wide-ranging dendritic arborization (Carp and Wolpaw 2001).

DISEASE AND TRAUMA OF LOWER MOTOR NEURONS

Dysfunction or degeneration of lower motor neurons can result in debilitating and even lethal paralysis. Amyotrophic lateral sclerosis, Kennedy's disease, progressive muscular atrophy, and spinal muscular atrophy are examples of hereditary and/or idiopathic diseases involving the degeneration of lower motor neurons. Myasthenia gravis is an example of an autoimmune disease involving

[3]Correspondence: bth@georgetown.edu

Cite this introduction as *Cold Spring Harb Protoc;* doi:10.1101/pdb.top070920

autoantibodies directed most commonly against acetylcholine receptors, which disrupts neurotransmission at the neuromuscular junction. Trauma to motor neurons induced by spinal cord injury, radiculopathy, or peripheral nerve injury can be transient or permanent depending on the severity and location of tissue crush or transection.

SPINAL MOTOR NEURONS IN CULTURE

Culturing of lower motor neurons is an important, albeit simplified, system for studying fundamental cellular and molecular neurobiological questions that may shed light on neurological diseases involving motor neurons. Cultures may be studied pure or nearly so, or mixed with other cells (distinctly purified or mixed), depending on the experimental paradigm. Such cultures have been used to study glial–neuronal interactions (Ullian et al. 2004), synaptogenesis (Ullian et al. 2004), axonal transport (Stommel et al. 2007), mitochondrial movement (Stommel et al. 2007), neurotrophin function (Hanson et al. 1998), and neuronal degeneration/apoptosis/necrosis (Hanson et al. 1998) by a handful of laboratories.

The accompanying protocol, Protocol 1: Purification and Culture of Spinal Motor Neurons from Rat Embryos (Graber and Harris), was derived from a methodology first reported by Camu and Henderson (1992) and later modified only slightly by Ben Barres' and our laboratories. It takes advantage of several distinct properties of rat lower motor neurons to isolate them away from their neighboring cells. First, an ideal stage in development after motor neurons are born (embryonic day 14 during rat gestation), but prior to extensive axonal extension or developmental apoptosis, is exploited (Yamamoto and Henderson 1999; Sendtner et al. 2000). Lower motor neurons cannot be viably isolated using this method after birth. After dissociating embryonic spinal cord tissue, which contains lower motor neurons among many other cell types, cells are separated based on cell density because motor neurons are uniquely large. Finally, this collected cell population is further purified based on selective immunopanning for motor neurons, which express the low-affinity nerve growth factor receptor often referred to as p75 (Yan and Johnson 1988). The near-pure lower motor neuron cultures are plated and seeded in defined conditions optimal for survival. These cultured motor neurons are rounded when initially plated to the growth substrate, and then project an elaborate array of axons and dendrites within five days.

REFERENCES

Altman J, Bayer SA. 1984. The development of the rat spinal cord. *Adv Anat Embryol Cell Biol* **85:** 1–164.

Camu W, Henderson CE. 1992. Purification of embryonic rat motoneurons by panning on a monoclonal antibody to the low-affinity NGF receptor. *J Neurosci Methods* **44:** 59–70.

Carp JS, Wolpaw JR. 2001. Motor neurons and spinal control of movement. *Encyclopedia of life sciences.* Wiley, Chichester, United Kingdom.

Graber DJ. Harris BT. 2013. Purification and culture of spinal motor neurons from rat embryos. *Cold Spring Harb Protoc* doi: 10.1101/pdb.prot074161.

Hanson MG Jr, Shen S, Wiemelt AP, McMorris FA, Barres BA. 1998. Cyclic AMP elevation is sufficient to promote the survival of spinal motor neurons in vitro. *J Neurosci* **18:** 7361–7371.

Sendtner M, Pei G, Beck M, Schweizer U, Wiese S. 2000. Developmental motoneuron cell death and neurotrophic factors. *Cell Tissue Res* **301:** 71–84.

Stommel EW, van Hoff RM, Graber DJ, Bercury KK, Langford GM, Harris BT. 2007. Tumor necrosis factor-α induces changes in mitochondrial cellular distribution in motor neurons. *Neuroscience* **146:** 1013–1019.

Ullian EM, Harris BT, Wu A, Chan JR, Barres BA. 2004. Schwann cells and astrocytes induce synapse formation by spinal motor neurons in culture. *Mol Cell Neurosci* **25:** 241–251.

Yamamoto Y, Henderson CE. 1999. Patterns of programmed cell death in populations of developing spinal motoneurons in chicken, mouse, and rat. *Dev Biol* **214:** 60–71.

Yan Q, Johnson EM Jr. 1988. An immunohistochemical study of the nerve growth factor receptor in developing rats. *J Neurosci* **8:** 3481–3498.

 Cite this introduction as *Cold Spring Harb Protoc*; doi:10.1101/pdb.top070920

Protocol 1

Purification and Culture of Spinal Motor Neurons from Rat Embryos

David J. Graber[1] and Brent T. Harris[2,3]

[1]Dartmouth Medical School, Department of Pathology, Lebanon, New Hampshire 03756; [2]Georgetown University Medical Center, Departments of Neurology and Pathology, Washington, D.C. 20057

We describe an immunopanning protocol to isolate, enrich, and culture spinal motor neurons from rat embryonic spinal cords. The method takes advantage of several distinct properties of rat lower motor neurons to isolate them from neighboring cells. First, an ideal stage in development after motor neurons are born (embryonic day 14 during rat gestation), but prior to extensive axonal extension or developmental apoptosis, is exploited. Lower motor neurons cannot be viably isolated using this method after birth. After dissociating embryonic spinal cord tissue, which contains lower motor neurons among many other cell types, the uniquely large motor neurons are enriched using density gradient centrifugation. Finally, the collected cell population is further purified based on selective immunopanning for motor neurons, which express the low-affinity nerve growth factor (NGF) receptor often referred to as p75. The near-pure lower motor neuron cultures are plated and seeded in defined conditions optimal for survival and can be maintained for several weeks. The expected yield is approximately 70,000 cells per embryonic spinal cord.

MATERIALS

It is essential that you consult the appropriate Material Safety Data Sheets and your institution's Environmental Health and Safety Office for proper handling of equipment and hazardous materials used in this protocol.

RECIPES: Please see the end of this protocol for recipes indicated by <R>. Additional recipes can be found online at http://cshprotocols.cshlp.org/site/recipes.

Reagents

Bovine serum albumin (BSA)

To prepare a stock of 4% BSA in Dulbecco's phosphate-buffered saline (D-PBS), dissolve 8 g of BSA (Sigma-Aldrich A4161) in 150 mL of D-PBS (HyClone SH30264.01) at 37°C. Adjust the pH to 7.4 with ~1 mL of 1 N NaOH. Bring the volume to 200 mL. Filter through a 0.22-µm filter. Store in 1-mL aliquots at −20°C.

Cytosine arabinoside (optional; see Step 31)

DNase

To prepare a 0.4% stock of DNase in Earle's balanced salt solution (EBSS), add 1 mL of EBSS (Sigma-Aldrich E6267) per 12,500 units of DNase (Worthington LS002007 or Sigma D4527). Keep on ice. Filter-sterilize, and store in 200-µL aliquots at −20°C.

[3]Correspondence: bth@georgetown.edu

Cite this protocol as *Cold Spring Harb Protoc*; doi:10.1101/pdb.prot074161

Fetal bovine serum (FBS) (Gibco/Life Technologies 26140)

Make 50-mL aliquots, then store at −20°C.

Goat anti-mouse IgG (Jackson ImmunoResearch 115-005-166)

Human placental laminin solution

Thaw human placental laminin (Sigma-Aldrich L6274) at 4°C. Make 10-µL aliquots and store at −20°C. Before use, solubilize 1.67 µg/mL of laminin in Neurobasal Medium (Gibco/Life Technologies 21103) with 100 U/mL penicillin and 100 mg/mL streptomycin.

Leibovitz's L-15 medium (Gibco/Life Technologies 11415-064)

Metrizamide solution

To prepare, dissolve 6.8 g of metrizamide (AK Scientific 69696) in 100 mL of PBS. Mix well. Filter through a 0.22-µm filter. Store in 5-mL aliquots at 4°C.

Alternatives to metrizamide solution include OptiPrep Density Gradient Medium (Sigma-Aldrich D1556) or Histodenz (Sigma-Aldrich D2158) (see Step 8).

Motor neuron growth medium, freshly prepared <R>

Mouse anti-p75 NGF receptor antibody (Abcam ab6172)

These antibodies are specific for rat p75. Alternatively, use undiluted supernatant from 192 hybridoma cells (see Step 4). Contact Ben Barres for hybridoma clone information.

PBS (calcium- and magnesium-free; Cellgro 21-031-CV)

Poly-D-lysine

To prepare a 1 mg/mL stock of poly-D-lysine, add 5 mL of water to a 5-mg bottle of poly-D-lysine (Sigma-Aldrich P6407). Filter through a 2-µm filter. Make 100-µL aliquots and store at −20°C.

Tris–HCl (100 mM, pH 7.4)

Dissolve 2.42 g of Trizma base in 200 mL dH_2O. Adjust the pH to 7.4 with HCl.

Trypan blue (Gibco/Life Technologies 15250-061)

Trypsin solution (TrypZean Solution; Sigma-Aldrich T3449)

Prepare 10-mL aliquots and store at −20°C.

Equipment

Centrifuge with rotor inserts for 15-mL conical tubes

Cell culture incubator

Optimal humified cell culture incubator settings are 37°C with 94% normal atmosphere and 6% CO_2.

CO_2 euthanization equipment for rats

Conical tubes (15 mL) (Corning 430055)

Coverslips (glass, 12-mm diameter; Carolina 633009), washed with ethanol

Dissection microscope

Fine-tip forceps (#5 or #55)

Hemacytometer

Inverted phase-contrast microscope

Polystyrene universal container (30 mL) (Lennox ABS200.05)

Scalpel with #10 blades

Scissors (small curved and very fine)

Sterile hood

Sterile Petri dishes (60- and 100-mm; Falcon 351007 and 351029, respectively)

Tissue culture plate (24-well) (Falcon 353047)

Water bath at 37°C

Cite this protocol as *Cold Spring Harb Protoc*; doi:10.1101/pdb.prot074161

METHOD

Figure 1 provides an overview of this 2-day procedure. Perform Steps 1 and 2 on the first day, and all subsequent steps on the second day. Conduct all steps at room temperature unless otherwise noted.

Prepare Cell Culture Growth Substrate and Immunopanning Dish

1. Coat coverslips with poly-D-lysine.

   i. Rinse ethanol-washed coverslips with sterile dH_2O four times in a Petri dish. Then place one coverslip in each well of a 24-well tissue culture plate.

   ii. Add 0.5 mL of sterile 20 µg/mL poly-D-lysine to each well and incubate for 30 min. Make sure that the coverslips are not floating and that they are completely submerged.

   iii. Discard the poly-D-lysine solution, and rinse the wells three times with sterile dH_2O, leaving dH_2O in the wells after the last rinse. Cover the tray, and store overnight.

   On the next day, proceed to Step 3.

2. Coat an immunopanning dish with secondary antibody.

   i. In a sterile hood, add 12 mL of goat anti-mouse IgG solution (3.75 µg/mL of goat anti-mouse IgG in 100 mM Tris–HCl [pH ~7.4]) to a 100-mm Petri dish.

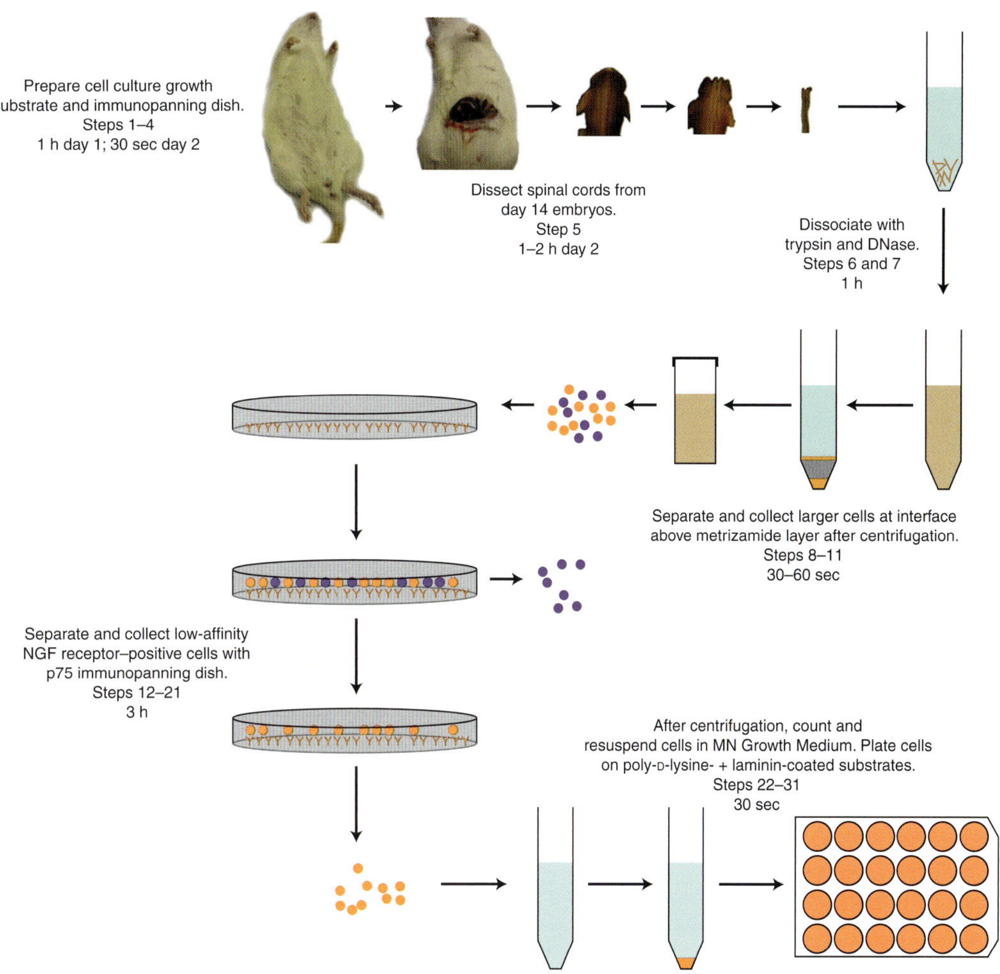

FIGURE 1. Isolation and enrichment of spinal motor neurons.

ii. Swirl the dish to cover the entire surface with antibody solution, cover, and store it at 4°C overnight on a level surface.

On the next day, proceed to Step 4.

3. Coat the coverslips with human placental laminin solution.

i. Discard the dH_2O from the wells, and add 0.3 mL of human placental laminin solution to each well. Make sure that the coverslips are not floating and they are completely submerged.

ii. Store the coverslips in the cell culture incubator until it is time to plate the cells.

4. Coat the immunopanning dish with primary antibody.

i. In a sterile hood, decant the anti-mouse IgG solution from the Petri dish and rinse three times with PBS.

ii. Add 10 mL of mouse anti-p75 NGF receptor antibody solution (1–10 µg/mL in PBS) to the Petri dish, cover, and set on a flat surface for at least 2 h.

Alternatively, use 10 mL of undiluted supernatant from a 192 hybridoma clone.

Dissect Embryonic Spinal Cords

This motor neuron purification procedure is specialized for derivation from embryonic day 14 rat spinal cords (day zero is the plug-check day after overnight mating).

5. Isolate embryonic spinal cords.

i. Euthanize a pregnant rat with CO_2. Rapidly dissect the lower abdomen and remove the uterus containing the embryos. Place the uterus into a sterile 100-mm Petri dish.

ii. Remove the embryos and embryonic sac from the uterus using fine-tip forceps and/or small surgical scissors. Place each embryo in a 60-mm Petri dish containing ice-cold sterile PBS.

iii. Decapitate each embryo at the infracranial notch just below the base of skull using fine-tip forceps or small curved scissors with the aid of a dissection microscope.

iv. Place the embryo ventral side down in the dish and position it so that the caudal end is to your nondominant hand. Use fine-tip forceps to remove the skin overlying the spinal cord, moving rostral to caudal.

v. Reposition the embryo with the rostral end of the spinal cord facing you. Gently use the forceps to work the enlarged portion of rostral spinal cord free from any remaining meninges or tissue, and obtain a clean tissue plane.

vi. With one set of forceps, pin down the embryo by puncturing at the level of each limb (taking care to avoid the spinal cord). Take hold of the rostral end of the spinal cord and, keeping the end of the cord steady, gently pull the body toward yourself from beneath to remove the cord from the body. Try to keep the entire cord intact.

vii. Carefully examine the spinal cord and use fine-tip forceps to remove any remaining meninges and/or dorsal root ganglia.

viii. Place the embryonic spinal cord into a Petri dish containing ice-cold sterile PBS. Repeat Steps 5.iii–5.vii for each embryo, and place the embryonic spinal cords into the same dish.

Dissociate Spinal Cord Tissue

6. Trypsinize the spinal cords.

i. In a sterile hood, transfer the spinal cords to a sterile 15-mL conical tube containing 2 mL of trypsin solution. Incubate for 15 min in a 37°C water bath, agitating every 3 min.

ii. Return the tube to the sterile hood, allow the tissue to settle, and discard as much trypsin solution as possible without losing tissue.

Cite this protocol as *Cold Spring Harb Protoc*; doi:10.1101/pdb.prot074161

7. DNase-treat and triterate the spinal cord tissue.
    i. Add 2 mL of Leibovitz's L-15 medium (L-15) supplemented with 0.2% BSA, 5% fetal bovine serum, and 0.02% DNase to the tube and agitate for 2–3 min.
    ii. Allow the tissue pieces to settle for 2 min and then collect the supernatant into a tube containing 4 mL of L-15. Layer this initial cell suspension slowly on top of a 1-mL cushion of 4% BSA. Centrifuge at 245*g* for 10 min.
    iii. During the centrifugation, add 2 mL of L-15 supplemented with 0.4% BSA and 0.004% DNase to the remaining cord tissue and slowly triturate two to three times until the suspension is cloudy. Allow the tissue pieces to settle for 2 min. After the tissue has settled, add the supernatant to a tube containing 1 mL of L-15.
    vi. Repeat Step 7.iii until no tissue pieces remain. Combine all cell suspensions into one tube.
    v. Retrieve the tube from the centrifuge, decant the supernatant, and resuspend the pellet in 1 mL of L-15. Add this cell suspension to rest of the cell suspensions collected in Step 7.vi.

Separate Cells by Density Gradient Centrifugation

8. Slowly layer the cell suspension from Step 7.v on top of a sterile 2-mL cushion of metrizamide solution in a 15-mL conical tube. Do not disturb the interface.

    Alternatively, use OptiPrep Density Gradient Medium or Histodenz instead of metrizamide. Concentrations, temperature, time, and centrifugal force must be optimized with these alternative solutions.

9. Centrifuge the layered tube at 515*g* for 15 min at 4°C with no brake.
10. Carefully collect the cell layer that contains the motor neurons (at the interface between the metrizamide and L-15 solutions). Add these cells to a tube containing 6 mL of L-15.
11. Slowly layer the collected cell suspension on top of a 1-mL cushion of 4% BSA. Centrifuge for 10 min at 245*g*.

Collect Lower Motor Neurons by Immunopanning

12. Discard the supernatant and resuspend the cell pellet in 10 mL of L-15. Transfer the cells to a 30-mL polystyrene universal container. Screw the cap on tightly and incubate for 90 min at 37°C.
13. Immediately before the end of this incubation, decant the antibody solution from the immunopanning Petri dish (from Step 4.ii), and rinse the dish three times with PBS.
14. Add the cell suspension to the panning Petri dish, cover, and incubate on a flat surface for 60 min.
15. After the incubation, swirl the Petri dish, and discard the supernatant.
16. Rinse the dish three times with PBS to remove loosely attached cells.
17. Add 4 mL of PBS and examine under an inverted phase-contrast microscope. Motor neurons are phase bright. If floating or loosely attached phase-dark cells remain, rinse with PBS two more times.
18. Discard the final PBS wash. Add 3 mL of trypsin solution, cover, and incubate at 37°C for 2–3 min to detach motor neurons from the Petri dish.
19. Add 7 mL of sterile 30% FBS in PBS, swirl, and then pipette the solution up and down across the entire surface of the Petri dish, washing the attached cells into the solution. Collect the cell suspension into a 15-mL conical tube.
20. Add another 3 mL of 30% FBS in PBS to the Petri dish. Collect the remaining cells as described in Step 19.
21. Rinse the plate with PBS, and examine it under an inverted phase-contrast microscope. If motor neurons remain attached to the dish, repeat Step 20 until all cells have been collected in the 15-mL conical tube.

Plate and Culture Lower Motor Neurons

22. Centrifuge the cell suspension for 10 min at 245g with no brake.
23. Decant the supernatant and resuspend the pellet in 1 mL of freshly prepared motor neuron growth medium.
24. Mix 10 µL of cell suspension with 10 µL of trypan blue, and count unstained cells on a hemacytometer using an inverted phase-contrast microscope. Determine the cell density.
25. Add an appropriate volume of motor neuron growth medium to the cell suspension to yield between 20,000 (low density) and 50,000 (high density) cells/mL.
26. Retrieve the tissue culture plate that contains the poly-D-lysine- and laminin-coated coverslips (from Step 3.ii).
27. Discard the laminin solution from the wells of the tissue culture plate.
28. Add 0.6 mL of cell suspension to each well.
29. Cover the plate and incubate it in the cell culture incubator.
30. Exchange 0.3 mL of culture supernatant with 0.3 mL of motor neuron growth medium every 3–4 d.
31. To reduce the proliferation of residual astrocytes in long-term cultures, add 100 nM of cytosine arabinoside to motor neuron cultures during days 4 and 8. (This does not noticeably harm the neurons.)

    Cultured motor neurons are rounded when initially plated onto the growth substrate, and then project an elaborate array of axons and dendrites within 5 d (Fig. 2). Lower motor neuron cell cultures can be maintained for several weeks.

RELATED INFORMATION

This protocol was derived from methodology first reported by Camu and Henderson (1992). It has been modified only slightly by Ben Barres' and our laboratories.

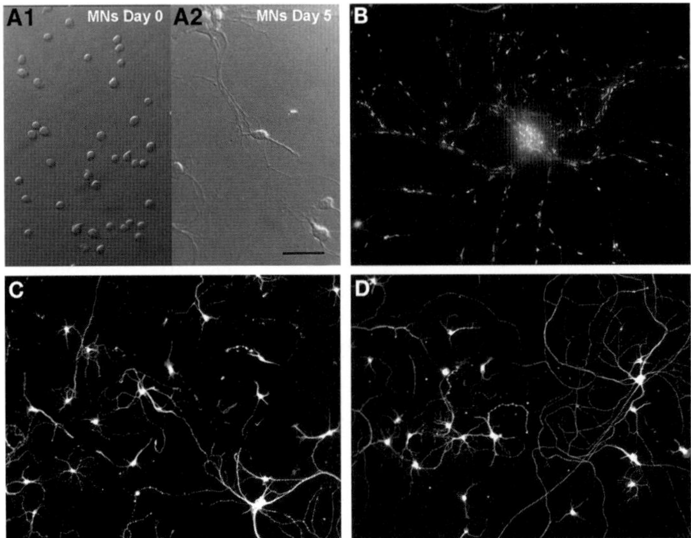

FIGURE 2. (*A*) Differential interference contrast (DIC) images of lower motor neurons at the day of plating (*A1*) and at 5 d in vitro (*A2*). (*B*) MitoTracker Green–labeled mitochondria in a cultured lower motor neuron (see also supplemental live cell movie in Stommel et al. 2007). (*C,D*) βIII-tubulin-labeled lower motor neurons derived from wild-type (*C*) or a combination of wild-type and G93A SOD1 transgenic embryos (*D*) after 4 d in vitro. (*B*, Reprinted from Stommel et al. 2007; *A,C,D*, courtesy of Brent Harris' laboratory.)

Cite this protocol as *Cold Spring Harb Protoc*; doi:10.1101/pdb.prot074161

RECIPES

BDNF Stock (10 μg/mL)

To prepare, dilute brain-derived neurotrophic factor (BDNF; Sigma-Aldrich B3795) to 10 μg/mL with sterile 0.2% BSA in Dulbecco's phosphate-buffered saline (D-PBS; HyClone SH30264.01). Make 20-μL aliquots, flash freeze in liquid nitrogen, and store at –80°C.

CNTF Stock (10 μg/mL)

To prepare, dilute ciliary neurotrophic factor (CNTF; Peprotech 450-13) to 10 μg/mL with sterile 0.2% BSA in Dulbecco's phosphate-buffered saline (D-PBS; HyClone SH30264.01). Make 20-μL aliquots, flash freeze in liquid nitrogen, and store at –80°C.

Forskolin Stock (4.2 mg/mL)

To prepare, add 1 mL of sterile DMSO to a 50-mg bottle of forskolin (Sigma-Aldrich F6886) and pipette up and down until the powder is fully resuspended. Transfer to a 15-mL conical tube and add an additional 11 mL of DMSO to achieve a final concentration of 4.2 mg/mL. Store in 20- and 80-μL aliquots at −20°C.

GDNF Stock (10 μg/mL)

To prepare, dilute glial-derived neurotrophic factor (GDNF; Peprotech 450-10) to 10 μg/mL with sterile 0.2% BSA in Dulbecco's phosphate-buffered saline (D-PBS; HyClone SH30264.01). Make 20-μL aliquots, flash freeze in liquid nitrogen, and store at −80°C.

Insulin Stock (0.5 mg/mL)

To 20 mL of sterile water, add 10 mg of insulin (Sigma-Aldrich I6634) and 100 μL of 1.0 N HCl. Mix well. Filter through a 0.22-μm filter. Store at 4°C for 4–6 wk.

Motor Neuron Growth Medium

1. Add Neurobasal Medium (Gibco/Life Technologies 21103) to the top of a sterile filter unit.
2. Add the following:

Reagent	Final concentration
B-27 Serum-Free Supplement (Gibco/Life Technologies 17504-044)	1×
SATO supplement (100×) <R>	1×
Insulin stock (0.5 mg/mL) <R>	5 μg/mL
Sodium pyruvate (Cambrex 13-115E or Gibco/Life Technologies 11360-070)	1 mM
L-Glutamine (Mediatech 25-005-CI)	2 mM
T3 stock (4 μg/mL) <R>	40 ng/mL
Mouse laminin (Gibco/Life Technologies 23017-015)	1 μg/mL
Isobutylmethylxanthine (RBI A-007)	2.2 μg/mL
Forskolin stock (4.2 mg/mL) <R>	417 ng/mL
NAC stock (5 mg/mL) <R>	5 μg/mL
Penicillin-streptomycin (Gibco/Life Technologies 15140-122)	100 U/mL (penicillin) 100 μg/mL (streptomycin)

Recently, some lots of B-27 supplement have been toxic; alternatives include B-27 minus antioxidants or NS21 supplement (R&D Systems AR008).

3. Filter-sterilize.
4. Add the following:

Reagent	Final concentration
BDNF stock (10 µg/mL) <R>	10 ng/mL
CNTF stock (10 µg/mL) <R>	10 ng/mL
GDNF stock (10 µg/mL) <R>	10 ng/mL

NAC Stock (5 mg/mL)

To prepare, dissolve 50 mg of *N*-acetyl-L-cysteine (NAC) powder (Sigma-Aldrich A8199) in 10 mL of Neurobasal Medium (Gibco/Life Technologies 21103). (The solution will be yellowish.) Filter through a 0.22-µm filter. Prepare 20- and 80-µL aliquots and store them at −20°C.

SATO Supplement (100×)

1. Prepare the following stock solutions (these should be made fresh; do not re-use):
   - Combine 5 mg of progesterone (Sigma-Aldrich P8783) and 200 µL of ethanol to make a progesterone stock solution.
   - Combine 4 mg of sodium selenite (Sigma-Aldrich S5261), 10 µL of 1 N NaOH, and 10 mL of Dulbecco's modified Eagle's medium (DMEM; Gibco/Life Technologies 11960-044) to make a sodium selenite stock solution.
2. Combine the following:

Reagent	Quantity (for 200 mL)	Final concentration (100×)
BSA (Sigma-Aldrich A4161)	2 g	10 mg/mL
Transferrin (Sigma-Aldrich T1147)	2 g	10 mg/mL
Putrescine (Sigma-Aldrich P5780)	320 mg	1.6 mg/mL
Progesterone stock solution	50 µL	6 µg/mL
Sodium selenite stock solution	2 mL	4 µg/mL

3. Bring to a total volume of 200 mL in DMEM, and then filter-sterilize. Aliquot and store at −20°C.

T3 Stock (4 µg/mL)

To prepare, dissolve 3.2 mg of 3,3′,5-triiodo-L-thyronine sodium salt (T3; Sigma-Aldrich T6397) in 400 µL of 0.1 N NaOH. Add 10 µL to 20 mL of Dulbecco's phosphate-buffered saline (D-PBS; HyClone SH30264.01). Filter through a 0.22-µm filter, discarding the first 10 mL. Make 200-µL aliquots, and store at −20°C.

REFERENCES

Camu W, Henderson CE. 1992. Purification of embryonic rat motoneurons by panning on a monoclonal antibody to the low-affinity NGF receptor. *J Neurosci Methods* **44:** 59–70.

Stommel EW, van Hoff RM, Graber DJ, Bercury KK, Langford GM, Harris BT. 2007. Tumor necrosis factor-α induces changes in mitochondrial cellular distribution in motor neurons. *Neuroscience* **146:** 1013–1019.

Cite this protocol as *Cold Spring Harb Protoc;* doi:10.1101/pdb.prot074161

CHAPTER 4

Purification and Culture of Dorsal Root Ganglion Neurons

J. Bradley Zuchero[1]

Department of Neurobiology, School of Medicine, Stanford University, Stanford, California 94305

Dorsal root ganglion neurons (DRGs) are sensory neurons that reside in ganglions on the dorsal root of the spinal cord. Here we introduce a method for the acute, prospective purification and culture of DRGs from rodents in a serum-free, defined medium, in the absence of glial cells. This immunopanning-based method facilitates the study of DRG biology and function.

INTRODUCTION

Dorsal root ganglion neurons (DRGs) are sensory neurons that occupy both the peripheral nervous system (PNS) and central nervous system (CNS). They reside in ganglions on the dorsal root of the spinal cord and have afferent axons that transmit sensory stimuli into the CNS. Multiple subsets of DRGs exist that respond to different sensory modalities (Dodd et al. 1984; Ruit et al. 1992; Friedel et al. 1997; Lallemend and Ernfors 2012). Their role in somatosensation has been extensively studied, and they have also been used at length to study neurite outgrowth, regeneration, and degeneration (Lindsay 1988; Wang et al. 2001; Teng and Tang 2006) and PNS and CNS myelination (Salzer and Bunge 1980; Wood and Bunge 1986; Chan et al. 2004).

NONPROSPECTIVE DRG PURIFICATION STRATEGIES

To date, purification strategies for DRG neurons do not prospectively isolate DRGs away from other contaminating cells, but generally rely on several weeks of growth in the presence of antimitotic substances (Wood 1976, 1980; Wood and Bunge 1986) or other cytotoxic treatments to target dividing cells and obtain pure cultures (Andersen et al. 2003). Typically these protocols require culturing neurons in the presence of serum, to which DRGs in the uninjured nerve are not normally exposed. Methods to separate DRGs away from other cell types based on their large size—by Percol gradient centrifugation or filtering dissociated cells through a 10-µm mesh to retain DRGs—result in a low yield (Goldenberg and De Boni 1983; Delree et al. 1989). An ideal purification strategy would (1) be prospective (meaning the cells are directly selected, without requiring extended culture time), (2) avoid prolonged exposure to cytoxic agents and serum, and (3) have a high yield.

PROSPECTIVE ISOLATION OF DRGS

In Protocol 1: Purification of Dorsal Root Ganglion Neurons from Rat by Immunopanning (Zuchero), we describe how to rapidly generate pure cultures of DRG neurons. These cultures are free of Schwann cells and other glia and thus can be used to study the role of glia in the biology of DRG neurons. They do not require extended time in the presence of antimitotic agents, nor do they require growth in the

[1]Correspondence: brad.zuchero@gmail.com

Cite this introduction as *Cold Spring Harb Protoc;* doi:10.1101/pdb.top073965

presence of serum. The protocol is based on previously described dissociation and neuronal immunopanning methods for other cell types (Huettner and Baughman 1986; Barres et al. 1988, 1992; Meyer-Franke et al. 1995) and uses defined medium with the B27-alternative NS21 (Chen et al. 2008). Immediately following purification, these DRGs require nerve growth factor (NGF) for survival, but to select for different populations of DRGs, NGF can be replaced with other neurotrophins (in isolation or in combination) the day after purification.

To enrich for DRGs, we first deplete blood cells and endothelial cells using their tight interaction with the lectin BSL-1. We then deplete glia using a monoclonal antibody to the tetraspanin protein CD9, which will recognize developing Schwann cells and oligodendrocyte precursor cells (Tole and Patterson 1993; Terada et al. 2002). It is worth noting that CD9 has been described as being expressed in DRGs (Tole and Patterson 1993), but in our hands we do not see significant binding of DRGs to our anti-CD9 immunopanning plates, perhaps because surface levels of CD9 are not as high in dissociated DRGs as in glia.

Peripheral CD9 expression increases after birth in rats (Kaprielian et al. 1995). Therefore, it should be possible to adapt this protocol to purification of neonatal or adult DRGs by modifying the dissociation protocol (Lindsay 1988; Delree et al. 1989; Malin et al. 2007). Additionally, CD9 is expressed ubiquitously by mouse and human glia (Nakamura et al. 1996; Terada et al. 2002; Sim et al. 2011), so this technique may be suitable for purifying DRGs from a wide range of species (Scott 1977).

REFERENCES

Andersen PL, Doucette JR, Nazarali AJ. 2003. A novel method of eliminating non-neuronal proliferating cells from cultures of mouse dorsal root ganglia. *Cell Mol Neurobiol* **23:** 205–210.

Barres BA, Silverstein BE, Corey DP, Chun LL. 1988. Immunological, morphological, and electrophysiological variation among retinal ganglion cells purified by panning. *Neuron* **1:** 791–803.

Barres B, Hart I, Coles H, Burne J, Voyvodic J, Richardson W, Raff M. 1992. Cell death and control of cell survival in the oligodendrocyte lineage. *Cell* **70:** 31–46.

Chan JR, Watkins TA, Cosgaya JM, Zhang C, Chen L, Reichardt LF, Shooter EM, Barres BA. 2004. NGF controls axonal receptivity to myelination by Schwann cells or oligodendrocytes. *Neuron* **43:** 183–191.

Chen Y, Stevens B, Chang J, Milbrandt J, Barres BA, Hell JW. 2008. NS21: Re-defined and modified supplement B27 for neuronal cultures. *J Neurosci Methods* **171:** 239–247.

Delree P, Leprince P, Schoenen J, Moonen G. 1989. Purification and culture of adult rat dorsal root ganglia neurons. *J Neurosci Res* **23:** 198–206.

Dodd J, Solter D, Jessell TM. 1984. Monoclonal antibodies against carbohydrate differentiation antigens identify subsets of primary sensory neurons. *Nature* **311:** 469–472.

Friedel RH, Schnürch H, Stubbusch J, Barde YA. 1997. Identification of genes differentially expressed by nerve growth factor- and neurotrophin-3-dependent sensory neurons. *Proc Natl Acad Sci* **94:** 12670–12675.

Goldenberg SS, De Boni U. 1983. Pure population of viable neurons from rabbit dorsal root ganglia, using gradients of Percoll. *J Neurobiol* **14:** 195–206.

Huettner JE, Baughman RW. 1986. Primary culture of identified neurons from the visual cortex of postnatal rats. *J Neurosci* **6:** 3044–3060.

Kaprielian Z, Cho KO, Hadjiargyrou M, Patterson PH. 1995. CD9, a major platelet cell surface glycoprotein, is a ROCA antigen and is expressed in the nervous system. *J Neurosci* **15:** 562–573.

Lallemend F, Ernfors P. 2012. Molecular interactions underlying the specification of sensory neurons. *Trends Neurosci* **35:** 373–381.

Lindsay RM. 1988. Nerve growth factors (NGF, BDNF) enhance axonal regeneration but are not required for survival of adult sensory neurons. *J Neurosci* **8:** 2394–2405.

Malin SA, Davis BM, Molliver DC. 2007. Production of dissociated sensory neuron cultures and considerations for their use in studying neuronal function and plasticity. *Nat Protoc* **2:** 152–160.

Meyer-Franke A, Kaplan M, Pfrieger F, Barres B. 1995. Characterization of the signaling interactions that promote the survival and growth of developing retinal ganglion cells in culture. *Neuron* **15:** 805–819.

Nakamura Y, Iwamoto R, Mekada E. 1996. Expression and distribution of CD9 in myelin of the central and peripheral nervous systems. *Am J Pathol* **149:** 575–583.

Ruit KG, Elliott JL, Osborne PA, Yan Q, Snider WD. 1992. Selective dependence of mammalian dorsal root ganglion neurons on nerve growth factor during embryonic development. *Neuron* **8:** 573–587.

Salzer JL, Bunge RP. 1980. Studies of Schwann cell proliferation. I. An analysis in tissue culture of proliferation during development, Wallerian degeneration, and direct injury. *J Cell Biol* **84:** 739–752.

Scott BS. 1977. Adult mouse dorsal root ganglia neurons in cell culture. *J Neurobiol* **18:** 417–427.

Sim FJ, McClain CR, Schanz SJ, Protack TL, Windrem MS, Goldman SA. 2011. CD140a identifies a population of highly myelinogenic, migration-competent and efficiently engrafting human oligodendrocyte progenitor cells. *Nat Biotechnol* **29:** 934–941.

Teng FY, Tang BL. 2006. Axonal regeneration in adult CNS neurons—Signaling molecules and pathways. *J Neurochem* **96:** 1501–1508.

Terada N, Baracskay K, Kinter M, Melrose S, Brophy PJ, Boucheix C, Bjartmar C, Kidd G, Trapp BD. 2002. The tetraspanin protein, CD9, is expressed by progenitor cells committed to oligodendrogenesis and is linked to β1 integrin, CD81, and Tspan-2. *Glia* **40:** 350–359.

Tole S, Patterson PH. 1993. Distribution of CD9 in the developing and mature rat nervous system. *Dev Dyn* **197:** 94–106.

Wang MS, Fang G, Culver DG, Davis AA, Rich MM, Glass JD. 2001. The WldS protein protects against axonal degeneration: A model of gene therapy for peripheral neuropathy. *Ann Neurol* **50:** 773–779.

Wood PM. 1976. Separation of functional Schwann cells and neurons from normal peripheral nerve tissue. *Brain Res* **115:** 361–375.

Wood PM, Bunge RP. 1986. Myelination of cultured dorsal root ganglion neurons by oligodendrocytes obtained from adult rats. *J Neurol Sci* **74:** 153–169.

Wood P, Okada E, Bunge R. 1980. The use of networks of dissociated rat dorsal root ganglion neurons to induce myelination by oligodendrocytes in culture. *Brain Res* **196:** 247–252.

Zuchero JB. 2014. Purification of dorsal root ganglion neurons from rat by immunopanning. *Cold Spring Harb Protoc* doi: 10.1101/pdb.prot074948.

 Cite this introduction as *Cold Spring Harb Protoc*; doi:10.1101/pdb.top073965

Protocol 1

Purification of Dorsal Root Ganglion Neurons from Rat by Immunopanning

J. Bradley Zuchero[1]

Department of Neurobiology, School of Medicine, Stanford University, Stanford, California 94305

Dorsal root ganglion neurons (DRGs) are sensory neurons that facilitate somatosensation and have been used to study neurite outgrowth, regeneration, and degeneration and PNS and CNS myelination. Studies of DRGs have relied on cell isolation strategies that generally involve extended culture in the presence of antimitotic agents or other cytotoxic treatments that target dividing cells. The surviving cells typically are dependent on serum for growth. Other methods, involving purification of DRGs based on their large size, produce low yield. In contrast, the immunopanning-based method described here for prospective isolation of DRGs from rodents allows for rapid purification in the absence of antimitotic agents and serum. These DRG cultures take place in a defined medium. They are free of Schwann cells and other glia and thus can be used to study the role of glia in the biology of DRG neurons.

MATERIALS

It is essential that you consult the appropriate Material Safety Data Sheets and your institution's Environmental Health and Safety Office for proper handling of equipment and hazardous materials used in this protocol.

RECIPES: Please see the end of this protocol for recipes indicated by <R>. Additional recipes can be found online at http://cshprotocols.cshlp.org/site/recipes.

Reagents

CD9 antibody, mouse anti-rat (BD Biosciences Pharmingen 551808)

Bovine serum albumin (BSA) (4%)

To prepare a stock of 4% BSA in Dulbecco's phosphate-buffered saline (D-PBS), dissolve 8 g of BSA (Sigma-Aldrich A4161) in 150 mL of D-PBS (Thermo Scientific HyClone SH3026401) at 37°C. Adjust the pH to 7.4 with ~1 mL of 1 N NaOH. Bring the volume to 200 mL. Filter through a 0.22-µm filter. Store in 1-mL aliquots at −20°C.

BSL-1 (*Griffonia simplicifolia* lectin; Vector Labs L-1100) (5 mg/mL)

Collagen I, rat tail (BD Biosciences Pharmingen 354236) (optional; see Step 1)

DNase (0.4%)

To prepare a 0.4% stock of DNase in Earle's balanced salt solution (EBSS), add 1 mL of EBSS (Ca-, Mg-free; Sigma-Aldrich E6267) per 12,500 units of DNase (Worthington LS002007). Keep on ice. Filter-sterilize, and store in 200-µL aliquots at −20°C.

---

[1]Correspondence: brad.zuchero@gmail.com

Cite this protocol as *Cold Spring Harb Protoc*; doi:10.1101/pdb.prot074948

Dulbecco's phosphate-buffered saline (D-PBS) with calcium and magnesium (e.g., Thermo Scientific HyClone SH3026401)

We add phenol red to the D-PBS to help ensure that all solutions are at neutral pH before using.

Ethanol (70%)
Fetal calf serum (FCS)

Make 50-mL aliquots of 100% FCS (Gibco/Life Technologies 10437-028) and heat-inactivate for 30 min at 55°C. Store the aliquots at −20°C.

5-Fluoro-2′-deoxyuridine (FUDR) (Sigma-Aldrich F0503) (for long-term culture; see Step 36)
Goat anti-mouse IgG + IgM (H + L) (Jackson ImmunoResearch 115-005-044)
Growth factors

Forskolin stock (4.2 mg/mL) <R>
Mouse nerve growth factor (NGF) (1 mg/mL stock; AbD Serotec PMP04Z)
BDNF stock (50 µg/mL) (optional; see Step 33) <R>
NT-3 stock (1 µg/mL) (optional; see Step 33) <R>

High-ovomucoid (high-ovo) stock solution (6×) <R>
Insulin (0.5 mg/mL) <R>
Laminin, mouse (1 mg/mL; Cultrex 3400-010-01)
L-Cysteine hydrochloride monohydrate (Sigma-Aldrich C7880)
Low-ovomucoid (low-ovo) stock solution (10×) <R>
Media

Neurobasal medium, filtered (Gibco 21103-049)
DRG base medium <R>
L15 dissection medium (Life Technologies 11415-064)

Combine 450 mL of Leibovitz's L15 medium with 50 mL of FCS. Filter-sterilize through a 0.22-µm filter and store at 4°C.

NaOH (1 M)
Papain (Worthington LS003126)
Phosphate-buffered saline (PBS) (e.g., Diamedix 1000-3)
Poly-D-lysine (PDL) stock (1 mg/mL) <R>
Rat, pregnant with E15 embryos, obtained on the second day of procedure
Tris-HCl (50 mM, pH 9.5)
Trypan blue (Life Technologies 15250-061)

Equipment

Aspirator
Bunsen burner
CO_2 chamber and tank for euthanizing rats
Conical tubes (15 and 50 mL)
Coverslips, glass, ethanol-washed
Dissection scissors, curved (e.g., ROBOZ RS-5675)
Forceps (#5, #55)
Hemacytometer
Hood, sterile for tissue culture work
Incubator at 37°C, 10% CO_2
Micropipettor (1 mL)
Microscopes, dissecting and phase contrast
Nitex mesh filter (20 µm) (Small Parts B0015H4H1A)

Cut the mesh into 4-inch squares and autoclave.

Pasteur pipettes

 Cite this protocol as *Cold Spring Harb Protoc*; doi:10.1101/pdb.prot074948

Petri dishes (6, 10, and 15 cm; Falcon or Nunc)
Pipettes, serological (2 and 10 mL)
Pipettor, powered (e.g., Pipet-Aid)
Razor blade, sterile
Syringe filter (0.22 µm)
Tabletop centrifuge (with 15 mL/50 mL centrifuge tube adaptors) at room temperature
Tissue culture plates, plastic (e.g., 24 well, Falcon, or NUNC)
Water bath, preset to 34°C

METHOD

This method begins with dissection of the rat embryos and creation of a single-cell suspension of DRGs. These cells then undergo several immunopanning steps before they are plated (see Fig. 1 for an overview). The entire protocol is a 2-d procedure; Steps 1and 2 are performed on the first day. Day 1 procedures (Steps 1 and 2) require ~1 h, plus a 4 h–overnight incubation for laminin coating. Day 2 procedures take 6–8 h. Perform all steps in a sterile tissue culture hood, with the exception of the dissection. All solutions in contact with cells must be sterile.

Preparation of Coverslips, Solutions, and Panning Dishes

1. Coat coverslips with PDL and laminin.

   i. Rinse ethanol-washed glass coverslips three times with sterile H_2O and dry completely (can be done in a Petri dish).

   ii. Dilute PDL stock 1:100 in sterile H_2O.

   iii. Cover each coverslip with 100 µL of diluted PDL and incubate for 30 min at room temperature.

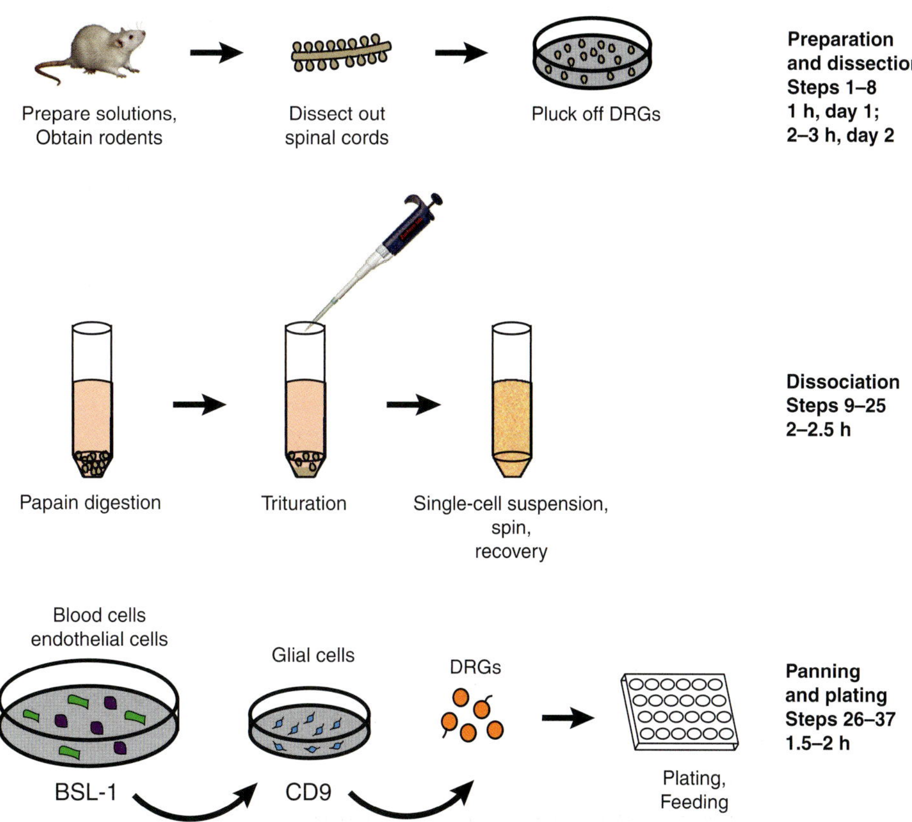

FIGURE 1. Immunopanning of DRGs by depleting glia with CD9 antibody.

iv. Rinse coverslips three times with sterile H_2O. During the third rinse, transfer the coverslips to tissue culture plates (e.g., 24-well plates).

v. Aspirate the H_2O and allow the coverslips to dry completely.

vi. Dilute the 1 mg/mL laminin 1:500 in filtered neurobasal medium.

vii. Cover each coverslip with 100 µL of diluted laminin and incubate for 4 h to 1 d at 37°C. If drops of laminin are disturbed and spill off the coverslips, add more laminin to cover.

Alternatively, to plate dense spots of DRGs (see Fig. 2 and Step 34), coat coverslips with a mixture of collagen and laminin as follows: After coating with poly-D-lysine and drying as in Steps 1.i–1.v, cover with laminin diluted 1:500 in 1:3 rat collagen:neurobasal medium. Incubate for 4–6 h at 37°C, and then aspirate the gelled collagen/laminin mixture thoroughly. Air-dry with the lids off in a sterile hood for at least 24 h, up to 4 d. Complete drying is essential for plating droplets to confine the DRG cell bodies to a small area.

2. Prepare panning dishes and incubate them overnight at 4°C.

- BSL1 dish: To a 15-cm Petri dish, add 20 mL PBS + 20 µL of 5 mg/mL BSL-1.
- CD9 dish: To a 10-cm Petri dish, add 10 mL of 50 mM Tris-HCl (pH 9.5) + 30 µL of goat-anti-mouse IgG + IgM (H + L).

Plates are initially hydrophobic, but after coating overnight, they become visibly hydrophilic. If plates are needed immediately, a quick but less preferable way to make them is to coat with secondary antibodies for 2 h at 37°C.

For preparing DRGs from two litters, prepare the CD9 dish by coating a 15-cm Petri dish with 20 mL of 50 mM Tris-HCl (pH 9.5) and 60 µL of goat-anti-mouse secondary antibody. The 15-cm BSL1 dish can be used for one or two litters.

3. Prepare solutions and tools.

i. Pour 10–20 mL of L15 dissection medium into a 15-cm dish and a 10-cm dish, and leave at room temperature for dissections.

ii. Pour ~3–4 mL of D-PBS into a 6-cm dish for dissections.

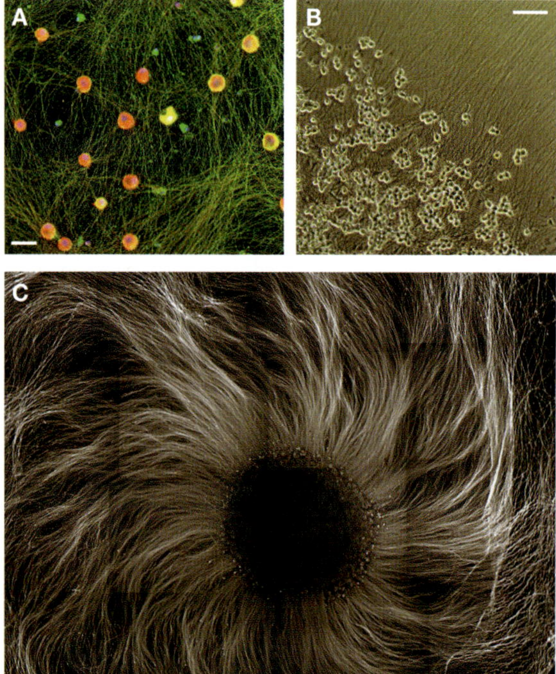

FIGURE 2. Example images of immunopanned and cultured DRG neurons. (*A*) DRGs plated in dispersed culture for 12 d, fixed and stained for axonal marker Tuj1 (green), pan-sodium channel (red), and nuclei (DAPI, blue). Scale bar, 20 µm. (*B*) DRG spots grown for 4 d, imaged in phase. Scale bar, 100 µm. (*C*) 40,000 DRGs plated in a dense spot were grown for 13 d, and fixed and stained for neurofilament to show the dense bed of axons. Scale bar, 500 µm.

Cite this protocol as *Cold Spring Harb Protoc*; doi:10.1101/pdb.prot074948

iii. In a 15-mL conical tube labelled "papain," combine 5 mL of D-PBS + 3 μL of 1 M NaOH, and warm in a 34°C water bath.

iv. Make 40 mL of 0.2% BSA by combining 38 mL of D-PBS + 2 mL of 4% BSA.

v. Warm and equilibrate DRG base medium in a 37°C, 10% CO_2 incubator, loosely capped to allow gas exchange.

vi. Sterilize the dissection area and tools by spraying with 70% ethanol.

4. Finish preparing the panning dishes by washing each dish from Step 2 three times with PBS and then incubating with primary antibody or blocking solution for ≥2 h at room temperature. Do not allow the coated dishes to dry.

- BSL-1 dish: 9 mL of 0.2% BSA from Step 3.iv.
- CD9 dish: 5 mL of 0.2% BSA + 15 μL of anti-rat CD9.

For preparing DRGs from two litters, coat a 15-cm CD9 dish (from Step 2) with 10 mL of 0.2% BSA + 30 μL of anti-rat CD9.

Dissection

5. Euthanize a timed pregnant rat (E15) with CO_2 according to appropriate animal protocol.
6. Dissect out the embryos.

i. Spray the abdomen with 70% ethanol, lift the skin above the stomach with forceps, and use a sterile razor to cut open the skin and muscle, taking care not to cut internal organs.

ii. Dissect out the placenta, containing the embryos, and pull into a 15-cm dish containing L15 dissection medium (prepared in Step 3).

iii. Using forceps and curved dissection scissors, carefully cut the placenta and enclosed amniotic sacs to pop out the embryos.

iv. Decapitate the embryos carefully without damaging their spinal cords.

v. Transfer the embryos to a 10-cm dish containing L15 dissection medium (prepared in Step 3).

7. Using the dissecting microscope, dissect out the spinal cords, one embryo at a time.

i. Cut off the tail.

ii. Cut off the ventral side (the belly) including the limbs; be careful to not cut too close to the spinal cord. Discard the ventral tissue.

iii. Use forceps to flip the embryo so that the ventral side faces up, and gently remove tissue, including the digestive track and blood vessels, until the rib cage is completely visible.

iv. Flip the embryo so that the dorsal side is facing up, with rostral side facing right.

v. Using forceps in your left hand, brace the embryo on either side of the spinal cord near the rostral side. Using forceps in your right hand, gently peel the skin above the spinal cord away from the bracing forceps.

vi. Move the bracing forceps (in your left hand) a few millimeters caudal, and repeat the gentle peeling and tearing of skin with the other forceps to begin to reveal the spinal cord and attached DRGs. Repeat until all of the dorsal skin has been removed from above the spinal cord.

vii. Flip the spinal cord 90° around its axis and use forceps in both hands to dissociate the spinal cord from the remaining ventral tissue (rib cage, etc.). Be careful not to strip the DRGs off of the spinal cord. This should result in a fully detached spinal cord with DRGs attached to its entire length. Set these aside in the dish until all the spinal cords are dissected.

8. Transfer spinal cords to a 6-cm Petri dish with D-PBS (prepared in Step 3), and again using the dissecting microscope, collect the DRGs.

i. Working down the length of each spinal cord, grasp the DRGs at their bases and pull off.

ii. Discard the stripped spinal cords.

iii. Bring the dish with the DRGs into the tissue culture hood and use a Pasteur pipette to transfer the DRGs into a 15-mL conical tube.

Swirling the dish to collect the DRGs in the center aids in collecting. Do not aspirate the DRGs past the flare of the Pasteur pipette, or it will be difficult to get them out.

Dissociation of DRGs

9. Allow the DRGs to settle, then aspirate the D-PBS. Wash the DRGs by adding 10 mL of D-PBS.

10. Allow the DRGs to resettle while you prepare the papain/DNase solution.

    i. Add 50 units of papain to the "papain" tube from Step 3.

    ii. Swirl the tube and return it to the 34°C water bath to dissolve the papain.

    iii. Measure 0.002 g of L-cysteine, add it to the papain tube, and bring the tube to the sterile hood.

    iv. Sterilize the papain solution by filtering it through a 0.22-µm filter into a new 15-mL conical tube.

    v. Add 50 µL of 0.4% DNase to the tube and invert to mix.

    For DRGs from two litters, increase the papain to 100 U and the DNase to 100 µL in 10 mL of D-PBS.

11. Without disturbing the settled DRGs, aspirate the D-PBS, and replace it with the papain/DNase solution.

12. Swirl the conical tube, incubate for 30 min in a 34°C water bath. Swirl the tube again halfway through the incubation to mix thoroughly.

13. During the incubation, make the following solutions:

    - Low-ovo inhibitor solution: 9 mL D-PBS + 1 mL 10× low-ovo stock solution + 100 µL of 0.4% DNase
    - High-ovo inhibitor solution: 5 mL D-PBS + 1 mL 6× high-ovo stock solution + ~3 µL of 1 M NaOH (to bring the solution to neutral pH)
    - Panning buffer: 18 mL D-BPS + 2 mL of 0.2% BSA + 200 µL of 0.5 mg/mL insulin

14. After the digestion in Step 12, wash the DRGs carefully.

    i. Bring the tube to the sterile hood and wait ~1 min for the tissue to settle.

    ii. Aspirate the excess liquid.

    iii. Gently add 4 mL of low-ovo inhibitor solution.

    iv. Wait for the tissue to settle; aspirate and discard the excess liquid.

15. Triturate the cells.

    i. Add 2 mL of low-ovo inhibitor solution to the DRGs.

    ii. Use a 1-mL micropipettor to triturate the cells up and down eight to 10 times, very slowly and gently.

    Be careful not to introduce bubbles, and to minimize the introduction of CO_2, avoid lifting the tip of the pipettor out of the solution. The low-ovo inhibitor solution will become cloudy.

    iii. Allow the tissue chunks to settle for 1–2 min.

    iv. Transfer the single cells (the cloudy solution on top of the chunks) to a new 50-mL conical tube. Avoid transferring any undissociated tissue.

    v. Add 1 mL of low-ovo inhibitor solution and repeat the trituration steps until all the chunks of tissue are gone. Each round of trituration can be gradually more forceful.

Cite this protocol as *Cold Spring Harb Protoc*; doi:10.1101/pdb.prot074948

vi. Use the rest of the low-ovo inhibitor solution to wash the remaining dissociated cells out of the tube and transfer to a 50-mL collection tube.

The goals of trituration are to (1) obtain isolated single cells and (2) subject those cells to as little manipulation as possible. It is essential to let the tissue chunks settle completely before collecting the single-cell suspension, but collect as much of this suspension as possible without disturbing the chunks (to avoid subjecting the single cells to more rounds of trituration). When first using this procedure, poor cell viability is common; with practice, however, improvements in viability should be seen. We typically see ≥95% viability at this step, with ~5–10 × 10^6 cells total.

16. Count the cells and assay viability by trypan blue exclusion.

    i. Combine 20 µL of cells with 80 µL of panning buffer or D-PBS.

    ii. Dilute this mixture 1:2 with trypan blue and count the cells using a hemacytometer.

    Expect ~5–10 × 10^6 cells per 10 to 12 embryos. This step can be done during the centrifugation in Step 19.

17. Use a 2-mL pipette to remove any chunks that have settled at the bottom of the 50-mL collection tube.

18. Carefully use a 10-mL pipette to slowly layer 6 mL of high-ovo inhibitor solution under the single-cell suspension.

    This step should lead to a clear layer of liquid beneath a cloudy cell suspension.

19. Pellet the cells through the high-ovo inhibitor solution by centrifuging for 10 min at 220*g* in a tabletop centrifuge at room temperature.

    We use a clinical centrifuge with ramp-up and ramp-down speeds set to moderate/low to prevent damage to cells. During the centrifugation, dissociated cells will descend through the high-ovo inhibitor solution, ensuring that complete inhibition of the papain occurs.

20. During the centrifugation in Step 19, make a cone from a Nitex filter.

    i. Use sterilized forceps (sprayed with ethanol and flamed with a Bunsen burner) to put a Nitex filter on top of a 50-mL conical tube.

    ii. Use a 1-mL micropipettor tip to depress the center of the filter into a cone (with the bottom of the cone inside the conical tube) while wetting with 1–2 mL of panning buffer.

21. After the centrifugation in Step 19, aspirate and discard the liquid, being careful to not disturb the pellet of cells.

22. Resuspend the cell pellet.

    i. Add 2 mL of panning buffer and gently pipette the solution and cells up and down with a 1-mL micropipettor.

    ii. Add 6 mL of panning buffer and mix gently.

23. Filter the solution, 1 mL at a time, through the Nitex mesh cone to separate the single cells from any remaining clumps of cells and chunks of tissue.

24. Wash the filter with 3 mL of panning buffer.

25. Incubate the tube in a 37°C, 10% CO_2 incubator for 30 min. Loosely cap the tube to allow gas exchange.

    This step allows antigens to return to the cell surface before immunopanning.

Panning

26. Rinse the BSL-1 panning dish three times with D-PBS, and pour off the final rinse.

27. Decant the cell suspension into the BSL-1 dish. Rinse the tube with 2–3 mL of panning buffer and add that buffer to the dish. Incubate the dish for 20 min at room temperature, gently shake, and then incubate for another 10 min (20 min total).

    It is important that the cells be allowed to settle onto the panning dish during this time. Periodic shaking ensures that all cells have access to the surface of the dish.

28. Rinse the CD9 dish nine times with D-PBS, and pour off the final rinse.

    Rinse the CD9 plate thoroughly, because CD9 antibody stock contains sodium azide.

29. Transfer the unbound cells from the BSL-1 dish to the CD9 dish.

    i. Shake the BSL-1 dish and decant it into the CD9 dish.

    ii. Prop the BSL-1 dish up at an angle and use a 1-mL micropipettor to transfer the ~1 mL of remaining cell suspension to the CD9 plate.

    iii. Incubate the CD9 plate for 15 min at room temperature, gently shake, and then incubate for another 15 min (30 min total).

30. Gently shake the CD9 plate, and decant the suspension of unbound cells into a 50-mL conical tube. Prop the dish up at an angle, and use a 1-mL micropipettor to transfer the ~1 mL of remaining cell suspension into the tube.

31. Count the cells and assay viability by trypan blue exclusion.

    i. Combine 20 µL of cells with 80 µL of panning buffer or D-PBS.

    ii. Dilute this mixture 1:2 with trypan blue, and count the cells using a hemacytometer.

    Expect ~1–2 × 10^6 cells per 10 to 12 embryos. This step can be done during the centrifugation in Step 32. We typically see ≥95% viability at this step. DRG neurons are easily distinguished by their large cell bodies and occasional axon stumps if dissociation was performed gently. We typically count only the DRG neurons and achieve >90% purity.

Plating

32. Pellet the cells by centrifuging at 220*g* in a tabletop centrifuge for 10 min at room temperature.

33. Add the following growth factors to the warmed DRG base medium from Step 3:

    - NGF at a 1:10,000 dilution (to 100 ng/mL)
    - Forskolin at a 1:1000 dilution (to 10 µM)
    - (*Optional*) BDNF at a 1:1000 dilution (to 50 ng/mL)
    - (*Optional*) NT-3 at a 1:1000 dilution (to 1 ng/mL)

    The presence of NGF immediately after the isolation procedure is critical for survival of DRGs, but it can be omitted after that time if other neurotrophins are present. For maximum survival, we typically grow DRGs in the presence of NGF, BDNF, and NT-3. If you wish to avoid growing DRGs in the presence of NGF (e.g., to establish myelinating cocultures with oligodendrocytes), replace the medium the next day with DRG base medium plus BDNF, NT-3, and forskolin (no NGF).

34. Aspirate the supernatant, resuspend the cell pellet in DRG base medium + growth factors, and plate the cells in a 24-well plate on the coverslips prepared in Step 1.

    i. To plate dispersed DRGs, resuspend the cells at 25,000–50,000 per 500 µL. Add the DRGs to the wells, and allow the plate to sit in the hood at room temperature for 5–10 min so the cells can settle and attach to the coverslips. Then gently transfer the plates to a 37°C, 10% CO_2 incubator.

    ii. To plate the DRGs in dense spots on dried collagen/laminin (see Fig. 2C), resuspend the cells at 20,000 DRGs per µL of medium. Plate 2–2.5 µL of cells in the center of dried coverslips, and allow the cells to settle for 20 min in the 37°C, 10% CO_2 incubator. The DRGs must not be allowed to dry out during this time. To prevent drying when plating in a 24-well plate, add 5–10 mL of D-PBS to the spaces between the wells, and carefully add 20–30 µL of medium around the edges of the coverslips. After 20 min, slowly add 500 µL of medium to the wells without flooding them (to avoid washing off the DRGs). To do so, pipette the medium in a circular motion around the outside of the well, allowing it to slowly close in around the central drop of DRGs.

Cite this protocol as *Cold Spring Harb Protoc*; doi:10.1101/pdb.prot074948

Feeding and Culturing DRGs

35. Feed DRGs by replacing half of their medium with fresh DRG base medium containing growth factors. To account for evaporation, remove 225 µL of medium from each well and replace it with 250 µL of fresh medium. Feed every 2–3 d.
36. For long-term culture, for the first feeding after the prep include 10 µM (final concentration) 5-fluoro-2′-deoxyuridine (FUDR) antimitotic in the medium to kill any contaminating dividing cells.
    i. Prepare the FUDR at 20 µM in DRG medium with growth factors and feed half volume.
    ii. The FUDR does not need to be washed out of the wells after use. Subsequent feedings do not normally require FUDR.
37. Monitor DRG axon growth carefully. DRGs will continue to extend axons for several weeks, although good coverage of the surface is typically achieved after 4–7 d of growth.

RELATED INFORMATION

For background information on this protocol, see Intoduction: Purification and Culture of Dorsal Root Ganglion Neurons (Zuchero).

RECIPES

BDNF Stock (50 µg/mL)

1. Prepare a master BDNF stock (1 mg/mL) by resuspending 1 mg of human brain-derived neurotrophic factor in powder form (BDNF; Peprotech 450-02) in 1 mL of cold, sterile 0.2% BSA (Sigma-Aldrich A-4161) that was prepared in D-PBS (Gibco 14287). Make 200-µL aliquots of the master stock, flash-freeze in liquid nitrogen, and store at −80°C.
2. To make a working BDNF stock, thaw a 200-µL aliquot of master stock on ice. At the same time, chill 3.8 mL of sterile 0.2% BSA solution on ice. Once chilled, add the master BDNF stock to the 0.2% BSA solution and mix well, but gently, to avoid foaming. Make 20-µL working aliquots and flash-freeze in liquid nitrogen. Store at −80°C.

DRG Base Medium

Reagent	Volume	Final concentration
DMEM (Gibco 11960-044)	37 mL	50%
Neurobasal (Gibco 21103-049)	37 mL	50%
Insulin (0.5 mg/mL) <R>	800 µL	5 µg/mL
Sodium pyruvate (Gibco 11360-070)	800 µL	1 mM
Penicillin/streptomycin (100×; Gibco 15140-122)	800 µL	1×
L-Glutamine (Gibco 25030-081)	800 µL	1 mM
SATO supplement, NB-based (100×) <R>	800 µL	1×
T3 stock (4 µg/mL) <R>	800 µL	40 ng/mL
NS21[a] (50×) <R>	1.6 mL	1×
NAC stock (5 mg/mL) <R>	80 µL	5 µg/mL
Final volume	80 mL	

Filter-sterilize the medium through a 0.22-µm filter and store at 4°C for up to 1 wk. Add growth factors (NGF, BDNF, NT-3, and forskolin) immediately before use.

[a]Alternatively, use N21-MAX Media Supplement (R&D Systems AR008) or B-27 (Invitrogen 17504-044). If using NS21 or B27 that contains T3, it is unnecessary to add T3 to DRG Base Medium.

Forskolin Stock (4.2 mg/mL)

To prepare, add 1 mL of sterile DMSO to a 50-mg bottle of forskolin (Sigma-Aldrich F6886) and pipette up and down until the powder is fully resuspended. Transfer to a 15-mL conical tube and add an additional 11 mL of DMSO to achieve a final concentration of 4.2 mg/mL. Store in 20- and 80-µL aliquots at −20°C.

High-Ovomucoid Stock Solution (6×)

BSA (Sigma-Aldrich A8806)
DPBS (Thermo Scientific HyClone SH3026401)
NaOH (1 N)
Trypsin inhibitor (Worthington LS003086)

1. Add 6 g of BSA to 150 mL D-PBS.
2. Add 6 g of trypsin inhibitor and mix to dissolve.
3. Add at least 1.5 mL of 1 N NaOH to adjust the pH; continue adding NaOH as necessary to bring up the pH to 7.4.
4. Bring the volume to 200 mL with D-PBS.
5. Filter-sterilize through a 0.22-µm filter.
6. Make 1.0-mL aliquots and store at −20°C.

Insulin (0.5 mg/mL)

1. Working under a sterile tissue culture hood, prerinse a 0.22-µm filter with sterile H_2O. Discard the flowthrough.
2. Add 50 mg of insulin (either recombinant human insulin [Sigma-Aldrich I2643] or bovine pancreas insulin [Sigma-Aldrich I6634]) to 100 mL of sterile H_2O in a 200-mL beaker. Immediately add 500 µL of 1 N HCl to adjust the pH.
3. Stir the mixture with a sterile pipette to dissolve the insulin completely; avoid creating bubbles.
4. Filter the solution through the pre-rinsed 0.22-µm filter.
5. Make 5-mL aliquots and store at 4°C for up to 4–6 wk. For use in media, dilute 100× to a concentration of 5 µg/mL.

Low-Ovomucoid Stock Solution (10×)

To prepare, add 3 g of BSA (Sigma-Aldrich A8806) to 150 mL D-PBS. Mix well. Add 3 g of trypsin inhibitor (Worthington LS003086) and mix to dissolve. Add ~1 mL of 1 N NaOH to adjust the pH to 7.4. Bring the volume to 200 mL with D-PBS. Filter-sterilize through a 0.22-µm filter. Make 1.0-mL aliquots and store at −20°C.

NAC Stock (5 mg/mL)

To prepare, dissolve 50 mg of *N*-acetyl-L-cysteine (NAC) powder (Sigma-Aldrich A8199) in 10 mL of Neurobasal Medium (Gibco/Life Technologies 21103). (The solution will be yellowish.) Filter through a 0.22-µm filter. Prepare 20- and 80-µL aliquots and store them at −20°C.

 Cite this protocol as *Cold Spring Harb Protoc*; doi:10.1101/pdb.prot074948

NS21 (50×) (Chem et al. 2008)

1. Add 12.5 g of BSA (Sigma-Aldrich A4161) to 100 mL of Neurobasal medium (Gibco 21103-049). Stir at room temperature for ~2 h to dissolve the BSA.
2. In a cold room at 4°C or in a beaker on ice, add each of the reagents in the table in the order that they are listed. Continue stirring the solution during the additions. Add all liquid stocks dropwise, and add all powders directly to the Neurobasal medium.
3. Aliquot the solution into 400-µL or 800-µL volumes and store at −30°C.
4. Dilute the NS21 stocks 1/50 in growth medium.

	Sigma-Aldrich catalog no.	Stock (mg/mL)	Final in 1× NS21 (µg/mL)	Amount stock add to 100 mL neurobasal	Notes on stock solution
Aqueous stocks (store at 4°C)					
L-Carnitine	C7518	2.0	2.0	5 mL	Resuspend all 10 mg in 5 mL dH_2O, add all of this
Ethanolamine	E0508	1.0	1.0	5 mL	Add 5 µL to 5 mL dH_2O, add all of this
D⁺-Galactose	G0625	15	15	5 mL	Weigh 75 mg and add to 5 mL dH_2O, add all of this
Putrescine	P5780	16.1	16.1	5 mL	Weigh 80 mg and add to 5 mL dH_2O, add all of this
Sodium selenite	S9133	0.01435	0.01435	5 mL	Resuspend all 1 mg in 70 mL dH_2O, add 5 mL of this
Ethanolic stocks (store under nitrogen at −70°C to minimize oxidation)					
Corticosterone	C2505	2.0	0.02	50 µL	
Linoleic acid	L1012	100	1.0	50 µL	
Linolenic acid	L2376	100	1.0	50 µL	
Lipoic acid (thioctic acid)		4.7	0.047	50 µL	
Progesterone		0.63	0.0063	50 µL	
Retinol acetate	R7882	10	0.1	50 µL	
Retinal, all trans (vit. A)		10	0.1	50 µL	
D,L-α-tocopherol (vit. E)	B5240	100	1.0	50 µL	
D,L-α-tocopherol acetate	T3001	100	1.0	50 µL	
Add fresh					
Albumin, bovine	A4161	100	2500	12.5 g	Weigh out 12.5 g and add directly to neurobasal, first step in the process
Catalase	C40	16	2.5	80 mg	Weigh out 80 mg and add directly
Glutathione (reduced)	G6013	1.0	1.0	5 mg	Weigh out 5 mg and add directly
Insulin	I6634	4.0	4.0	5 mL	Weigh out 20 mg, dissolve in 5 mL of 0.1% acetic acid, add all of this
Superoxidase dismutase	S5395	2.5	2.5	12.5 mg	Calculate 12.5 mg from weight on three bottles and add directly (mix of new and old lots)
Transferrin	T5391	2.0	5.0	10 mg	Add all 10 mg directly (rinse bottle with the B27 mix)

Continued

	Sigma-Aldrich catalog no.	Stock (mg/mL)	Final in 1× NS21 (µg/mL)	Amount stock add to 100 mL neurobasal	Notes on stock solution
T3 (triiodol-L-thyronine)	T6397	0.002	0.002	5 µL	Weigh out 10 mg, dissolve in 5 mL of 0.1 M NaOH, add all of this[a]

[a]We typically omit T3 from the recipe to enable making media lacking T3 for some cell types. If T3 is omitted, it will need to be added to DRG Base Medium separately.

NT-3 Stock (1 µg/mL)

1. Prepare a master stock of neurotrophin-3 (NT-3; Peprotech 450-03) at 1–0.1 mg/mL according to manufacturer's instructions. (Instructions may vary from lot to lot; generally, NT-3 is dissolved in buffer [e.g., Dulbecco's phosphate-buffered saline] plus 0.2% BSA.) Store at −80°C.
2. Prepare 0.2% BSA (Sigma-Aldrich A4161) in Dulbecco's phosphate-buffered saline (D-PBS; HyClone SH30264.01). Filter-sterilize and chill.
3. Dilute an aliquot of NT-3 master stock to 1 µg/mL in sterile, chilled 0.2% BSA prepared in Step 2.
4. Aliquot the NT-3 working stock from Step 3 (e.g., 20 µL/tube) and snap-freeze in liquid nitrogen.
5. Store aliquots at −80°C.

Poly-D-Lysine (PDL) Stock (1 mg/mL)

1. Resuspend poly-D-lysine (PDL; Sigma-Aldrich P6407; molecular weight 70–150 kDa) at 1 mg/mL in borate buffer by combining 50 mg of PDL and 50 mL of 0.15 M boric acid (pH 8.4).
2. Filter to sterilize, and then aliquot (e.g., 100 µL/tube). Store aliquots at −20°C.

SATO Supplement, NB-Based (100×)

1. Prepare the following stock solutions (these should be made fresh; do not reuse):
   - Combine 2.5 mg of progesterone (Sigma-Aldrich P8783) and 100 µL of ethanol to make a progesterone stock solution.
   - Combine 4.0 mg of sodium selenite (Sigma-Aldrich S5261), 10 µL of 1 N NaOH, and 10 mL of Neurobasal (NB, Gibco 21103-049) to make a sodium selenite stock solution.
2. Add the following to 80 mL of Neurobasal medium:

Reagent	Quantity	Final concentration in medium (1×)
BSA (Sigma-Aldrich A4161)	800 mg	100 µg/mL
Transferrin (Sigma-Aldrich T1147)	800 mg	100 µg/mL
Putrescine dihydrochloride (Sigma-Aldrich P5780)	128 mg	16 µg /mL
Progesterone stock solution	20 µL	60 ng/mL (0.2 µM)
Sodium selenite stock solution	800 µL	40 ng/mL

3. Mix well, and filter-sterilize through a prerinsed 0.22-µm filter. Make 200-µL or 800-µL aliquots, and store at −20°C.

 Cite this protocol as *Cold Spring Harb Protoc*; doi:10.1101/pdb.prot074948

T3 Stock (4 μg/mL)

To prepare, dissolve 3.2 mg of 3,3′,5-triiodo-L-thyronine sodium salt (T3; Sigma-Aldrich T6397) in 400 μL of 0.1 N NaOH. Add 10 μL to 20 mL of Dulbecco's phosphate-buffered saline (D-PBS; HyClone SH30264.01). Filter through a 0.22-μm filter, discarding the first 10 mL. Make 200-μL aliquots, and store at −20°C.

ACKNOWLEDGMENTS

We thank Jonah Chan, Ye Zhang, Amanda Brosius Lutz, Jack Wang, and Arun Thottumkara for helpful discussions and advice and Adiljan Ibrahim for excellent technical assistance. J.B.Z. is a Howard Hughes Medical Institute Fellow of the Life Sciences Research Foundation. This protocol was developed in Ben Barres' laboratory. We thank the Myelin Repair Foundation and the National Institutes of Health (NIH) for grant R01 EY10257 for support.

REFERENCES

Chen Y, Stevens B, Chang J, Milbrandt J, Barres BA, Hell JW. 2008. NS21: Re-defined and modified supplement B27 for neuronal cultures. *J Neurosci Methods* 171: 239–247.

Zuchero JB. 2014. Purification and culture of dorsal root ganglion neurons. *Cold Spring Harb Protoc* doi: 10.1101/pdb.top073965.

CHAPTER 5

Purification and Culture of Astrocytes

Lynette C. Foo[1]

Department of Neurobiology, School of Medicine, Stanford University, Stanford, California 94305

For years, studies of neuronal–glial interactions have relied on the use of astrocytes derived from the extended culture of immature precursor cells isolated from the neonatal rodent brain. Although the astrocytes cultured under these selective cell survival conditions have been important tools for understanding astrocyte behavior, they do not necessarily reflect the behavior and function of mature astrocytes. We have developed methods for acute, prospective isolation and culture of mature astrocytes from rodent brains in a serum-free, defined medium. These immunopanning-based methods facilitate the study of astrocyte biology and function.

STANDARD ASTROCYTE PREPARATIONS AND THEIR LIMITATIONS

Astrocytes constitute nearly half the cells of the human brain, but their functions have yet to be fully elucidated. They have long been known to play passive roles in the nervous system—for instance, by providing trophic support for neurons (Allen and Barres 2009). However, recent evidence has pointed to important active roles for astrocytes, including in the control of synapse formation, function, and plasticity in the developing, normal, and diseased brain (Eroglu and Barres 2010; Sofroniew and Vinters 2010).

A major limitation in studies of neuronal–glial interactions is that purification and culture methods have not yet been developed for mature astrocytes. The procedure developed by McCarthy and de Vellis (1980) involving the preparation of astrocytes from the neonatal rodent brain has long served as an in vitro proxy for astrocytes; we term astrocytes prepared in this manner "MD-astrocytes." Although these cultures have increased our understanding of astrocytic function, they have several drawbacks. These cultures select for a small population of unidentified precursor cells (McCarthy and de Vellis 1980; Hildebrand et al. 1997) that express several astrocytic markers, but appear to have an immature or reactive phenotype (Cahoy et al. 2008). Much of what we know about astrocytes derives from studies of these immature preparations; these studies have suggested important roles for astrocytes in neuronal survival (Banker 1980) and synaptogenesis (Eroglu and Barres 2010).

Standard methods for preparing MD-astrocytes begin by creating a cell suspension from P0–P1 neonatal mouse or rat cortex (McCarthy and de Vellis 1980). Small modifications of this method, such as preparing the single-cell suspension without the use of proteases, have been reported (Schwartz and Wilson 1992; Wu and Schwartz 1998), but all reported methods rely on the same general principle of selective cell survival. It has long been held that neurons are irreversibly postmitotic and require survival signals, whereas astrocytes are thought to retain the ability to divide and to be independent of these signals. Thus, for MD-astrocyte preparations, single-cell suspensions are plated at high density in serum-containing medium, and most cells, including neurons, die, but a small percentage of cells

[1]Correspondence: lynettefoo@gmail.com

Cite this introduction as *Cold Spring Harb Protoc*; doi:10.1101/pdb.top070912

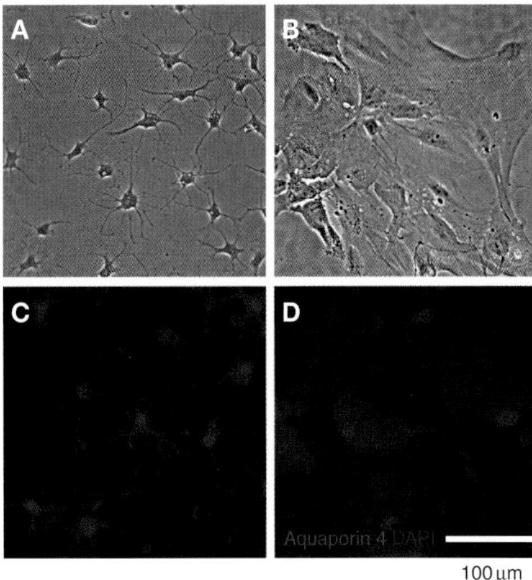

FIGURE 1. Phase (*A,B*) and fluorescent (*C,D*) images of IP- (*A,C*) and MD- (*B,D*) astrocyte cultures cultured for 14 d in vitro (DIV). IP-astrocytes are process-bearing, whereas MD-astrocytes are flat and fibroblast-like.

survive and proliferate. The cells eventually stratify into two main populations, the astrocytes and oligodendrocytes. The oligodendrocytes grow on top of the astrocytes and can be shaken off, leaving behind the MD-astrocytes. It is important to recognize that this method selects for cells in neonatal animals that can survive and proliferate under these culture conditions; the method is not prospective. Prospective isolation refers to the direct selection of cells without the use of steps that extend over a long period of time, and it ensures that a representative population of cells is isolated from the whole cell suspension.

PROSPECTIVE ISOLATION OF ASTROCYTES

The purification methods we have developed allow prospective, acute isolation of astrocytes from postnatal brain after they have achieved a relatively mature phenotype (Cahoy et al. 2008; Foo et al. 2011). The immunopanning-based method for purification and culture of mature astrocytes is described in Protocol 1: Purification of Rat and Mouse Astrocytes by Immunopanning (Foo [a]). We also employ a method that combines immunopanning with fluorescence-activated cell sorting (FACS) to isolate enhanced green fluorescent (eGFP) protein under astrocyte promoters in transgenic mice (see Protocol 2: Purification of Astrocytes from Transgenic Rodents by Fluorescence-Activated Cell Sorting [Foo (b)]). We call the cells isolated by these methods "IP-astrocytes" (immunopanned astrocytes). Unlike MD-astrocytes, astrocytes purified and cultured by immunopanning retain gene profiles in serum-free culture that closely resemble their in vivo gene profiles (see Fig. 1). IP-astrocytes still strongly promote neuronal survival as well as synapse formation and function, but significant differences between MD-astrocytes and acutely isolated astrocytes suggest that MD-astrocytes are not a good culture system for studying the function of astrocytes (Foo et al. 2011). Use of IP-astrocytes has led to the discovery of several important and fundamental properties of astrocytes. For instance, although it was previously believed that astrocytes do not need survival factors, we discovered that astrocytes do require trophic support for survival, and we identified a trophic factor that confers astrocytes' survival in vitro (Foo et al. 2011). Through gene expression studies of acutely isolated astrocytes from postnatal mice, we found that several phagocytic pathways are highly expressed in astrocytes, suggesting active phagocytic roles of astrocytes in the brain (Cahoy et al. 2008). These results indicate the importance of immunopanning-based purification methods for the study of astrocyte biology and function.

 Cite this introduction as *Cold Spring Harb Protoc*; doi:10.1101/pdb.top070912

ACKNOWLEDGMENTS

Our IP-astrocyte purification procedures are based on previously described dissociation (Huettner and Baughman 1986; Segal et al. 1998) and immunopanning purification protocols for other cell types (Barres et al. 1988, 1992; Meyer-Franke et al. 1995). We thank Dr. Jim Huettner for helpful comments on the adaptation of his neuronal dissociation procedure to allow for glial purification.

REFERENCES

Allen N, Barres B. 2009. Neuroscience: Glia—More than just brain glue. *Nature* **457:** 675–677.

Banker G. 1980. Trophic interactions between astroglial cells and hippocampal neurons in culture. *Science* **209:** 809–810.

Barres BA, Silverstein BE, Corey DP, Chun LL. 1988. Immunological, morphological, and electrophysiological variation among retinal ganglion cells purified by panning. *Neuron* **1:** 791–803.

Barres B, Hart I, Coles H, Burne J, Voyvodic J, Richardson W, Raff M. 1992. Cell death and control of cell survival in the oligodendrocyte lineage. *Cell* **70:** 31–46.

Cahoy JD, Emery B, Kaushal A, Foo LC, Zamanian JL, Christopherson KS, Xing Y, Lubischer JL, Krieg PA, Krupenko SA, et al. 2008. A transcriptome database for astrocytes, neurons, and oligodendrocytes: A new resource for understanding brain development and function. *J Neurosci* **28:** 264–278.

Eroglu C, Barres BA. 2010. Regulation of synaptic connectivity by glia. *Nature* **468:** 223–231.

Foo LC. 2013a. Purification of rat and mouse astrocytes by immunopanning. *Cold Spring Harb Protoc* doi: 10.1101/pdb.prot074211.

Foo LC. 2013b. Purification of astrocytes from transgenic rodents by fluorescence-activated cell sorting. *Cold Spring Harb Protoc* doi: 10.1101/pdb.prot074229.

Foo LC, Allen NJ, Bushong EA, Ventura PB, Chung W-S, Zhou L, Cahoy JD, Daneman R, Zong H, Ellisman MH, et al. 2011. Development of a method for the purification and culture of rodent astrocytes. *Neuron* **71:** 799–811.

Hildebrand B, Olenik C, Meyer DK. 1997. Neurons are generated in confluent astroglial cultures of rat neonatal neocortex. *Neuroscience* **78:** 957–966.

Huettner JE, Baughman RW. 1986. Primary culture of identified neurons from the visual cortex of postnatal rats. *J Neurosci* **6:** 3044–3060.

McCarthy KD, de Vellis J. 1980. Preparation of separate astroglial and oligodendroglial cell cultures from rat cerebral tissue. *J Cell Biol* **85:** 890–902.

Meyer-Franke A, Kaplan M, Pfrieger F, Barres B. 1995. Characterization of the signaling interactions that promote the survival and growth of developing retinal ganglion cells in culture. *Neuron* **15:** 805–819.

Schwartz JP, Wilson DJ. 1992. Preparation and characterization of type 1 astrocytes cultured from adult rat cortex, cerebellum, and striatum. *Glia* **5:** 75–80.

Segal M, Baughman R, Jones K. 2010. Stanford WebLogin (Culturing nerve cells).

Sofroniew M, Vinters H. 2010. Astrocytes: Biology and pathology. *Acta Neuropathol* **119:** 7–35.

Wu VW, Schwartz JP. 1998. Cell culture models for reactive gliosis: New perspectives. *J Neurosci Res* **51:** 675–681.

Protocol 1

Purification of Rat and Mouse Astrocytes by Immunopanning

Lynette C. Foo[1]

Stanford University, School of Medicine, Department of Neurobiology, Stanford, California 94305

We describe the use of immunopanning to purify rodent astrocytes. Immunopanning of astrocytes permits the prospective isolation of astrocytes from the rodent brain. Prospective isolation refers to the direct selection of cells without multiple steps that extend over a few days, thereby permitting the selection of a representative proportion of the astrocytes from the cortex. Because immunopanning is a very gentle process, the cells are viable at the end of the preparation and can be cultured in a serum-free medium containing heparin-binding EGF-like growth factor (HBEGF), the critical survival factor of astrocytes in vitro, for at least 2 wk. This protocol was initially established and verified with rats, but modifications for the purification of mouse cells are also described here.

MATERIALS

It is essential that you consult the appropriate Material Safety Data Sheets and your institution's Environmental Health and Safety Office for proper handling of equipment and hazardous materials used in this protocol.

RECIPES: Please see the end of this protocol for recipes indicated by <R>. Additional recipes can be found online at http://cshprotocols.cshlp.org/site/recipes.

Reagents

Animals (rats or mice) for dissection

Brain dissections are typically done on animals between P0 and P10 but can be done up to P15. Cell yield and viability decrease dramatically with age, because it is more difficult to dissociate older tissue with increased myelination in the brain and increased process elaboration of the astrocytes. Hence, there is more damage to the astrocytes with older animals.

BSA (4%)

To prepare a stock of 4% BSA in Dulbecco's phosphate-buffered saline (D-PBS), dissolve 8 g of BSA (Sigma-Aldrich A4161) in 150 mL of D-PBS (HyClone SH30264.01) at 37°C. Adjust the pH to 7.4 with ~1 mL of 1 N NaOH. Bring the volume to 200 mL. Filter through a 0.22-µm filter. Store in 1-mL aliquots at −20°C.

BSL-1 (*Griffonia simplicifolia* lectin; Vector Labs L-1100; 5 mg/mL)

DNase (0.4%)

To prepare a 0.4% stock of DNase in Earle's balanced salt solution (EBSS), add 1 mL of EBSS (Sigma-Aldrich E6267) per 12,500 units of DNase (Worthington LS002007 or Sigma-Aldrich D4527). Keep on ice. Filter-sterilize, and store in 200-µL aliquots at −20°C.

Dulbecco's phosphate-buffered saline (D-PBS) containing Mg^{2+} and Ca^{2+} (HyClone SH 30264.01)

Store at room temperature.

[1]Correspondence: lynettefoo@gmail.com

Cite this protocol as *Cold Spring Harb Protoc*; doi:10.1101/pdb.prot074211

Earle's balanced salt solution (EBSS without Mg^{2+} or Ca^{2+}, 1×; Sigma-Aldrich E6267), equilibrated as in Step 2.ii

Store at room temperature until opened, and then store at 4°C.

Enzyme stock solution <R>

Fetal calf serum (FCS)

Dilute 100% FCS (Gibco/Life Technologies 10437-028) in a 50/50 mixture of DMEM (Gibco/Life Technologies 11960-044) and Neurobasal (Gibco/Life Technologies 21103-049). The 100% FCS should be heat-inactivated for 30 min at 55°C and stored at −20°C. Dilute to 30% before use.

High-ovomucoid (high-ovo) stock solution (10×) <R>

Inhibitor stock solution <R>

L-Cysteine hydrochloride monohydrate (Sigma-Aldrich C7880)

Low-ovomucoid (low-ovo) stock solution (10×) <R>

Media (equilibrate as in Step 2.ii before use)

IP-astrocyte base medium <R>
IP-astrocyte base medium containing 5 ng/mL HBEGF (Sigma-Aldrich E4643)

Papain (Worthington LS003126)

Primary antibodies

Mouse anti-human integrin β5 (ITGB5), affinity purified (eBioscience 14-0497-82)
O4 hybridoma supernatant mouse IgM (Bansal et al. 1989)
Rat anti-mouse CD45 (BD Biosciences Pharmingen 550539)

The antibodies listed here are for use with the rat immunopanning protocol. See the end of the Method section for the antibodies needed for mouse immunopanning.

Secondary antibodies

Goat anti-mouse IgG + IgM (H + L) (Jackson ImmunoResearch 115-005-044)
Goat anti-mouse IgM μ-chain specific (Jackson ImmunoResearch 115-005-020)
Goat anti-rat IgG (H + L) (Jackson ImmunoResearch 112-005-167)

The antibodies listed here are for use with the rat immunopanning protocol. See the end of the Method section for the antibodies needed for mouse immunopanning.

Tris-HCl (50 mM, pH 9.5), sterilized

Trypan blue

Trypsin (30,000 units/mL) (Sigma-Aldrich T9935)

Do not store aliquoted trypsin at temperatures warmer than −80°C. Warmer temperatures allow a low level of trypsin activity, enabling the enzyme to cleave itself and render itself inactive.

Equipment

ACLAR plastic coverslips, ethanol-washed and PDL-coated

In practice, we coat ethanol-washed ACLAR coverslips with poly-D-lysine (PDL) after adding them individually to each well of a 24-well plate (see Step 56). ACLAR is very hydrophobic; add 500 μL of 0.01 mg/mL PDL (prepared in sterile H_2O) to cover each coverslip completely. If the coverslip floats up, tap it down with a pipette tip. Coat the coverslips with PDL for at least 30 min. Wash the coverslips in the wells three times with sterile H_2O before adding medium.

Aspirator

Conical tubes (50 mL)

Dissection equipment

Decapitation scissors, large (ROBOZ RS-6820)
Dissecting microscope

Dissection scissors, curved (ROBOZ RS-5675)
Forceps, #5
Forceps, curved
Scalpel with No.10 scalpel blade
Scissors, medium-sized (ROBOZ RS-6702)
Scoop, perforated (Moria MC17)

Equilibration setup (a source of carbon dioxide [5% CO_2/95% O_2] with a line leading to a sterile hood)

Heat block preset to 34°C in a sterile hood

Hemacytometer

Incubator at 37°C, 10% CO_2

Microscope, phase contrast

Nitex mesh filter (Tetko HC3-20)

Cut into 3-inch squares, wrap in small packets of foil, and autoclave.

Petri dishes (6 cm, 15 cm)

Petri dish lids (6 cm) with a hole in the center that accommodates a 0.22-µm filter

Use flamed forceps (spray with ethanol and use a Bunsen flame to sterilize) to melt the hole into the center of the lid.

Pipette (1-mL)

Pipettes, serological (2-mL, 5-mL, 10-mL)

Pipettor, powered (e.g., Pipet-Aid)

Refrigerator preset to 4°C

Sterile hood for tissue culture

Syringe, 20 mL

Syringe filters (0.22 µm)

Tabletop centrifuge (with 15 mL/50 mL centrifuge tube adaptors) at room temperature

Tissue culture plates (15-cm)

Tissue culture plates, multiwell (24-well)

Water bath at 34°C

METHOD

This method describes the purification of rat astrocytes (see Fig. 1 for an overview). It is a 2-d procedure; Step 1 is performed on the first day. Modifications required for purification of mouse astrocytes are outlined at the end.

Preparation of Panning Dishes

1. In a sterile hood, prepare the following panning dishes by coating six 15-cm Petri dishes with 25 mL of 50 mM Tris-HCl (pH 9.5) per dish and the appropriate secondary antibody solution:
    - One "Secondary antibody only" dish: 60 µL goat anti-mouse IgG + IgM (H + L)
    - One "BSL-1" dish: 40 µL 50 mM Tris–HCl (pH 9.5) + 20 µL BSL-1
    - One "CD45" dish: 60 µL goat anti-rat IgG (H + L)
    - Two "O4" dishes: 60 µL goat anti-mouse IgM µ-chain specific
    - One "ITGB5" dish: 60 µL goat anti-mouse IgG + IgM (H + L)

Cite this protocol as *Cold Spring Harb Protoc*; doi:10.1101/pdb.prot074211

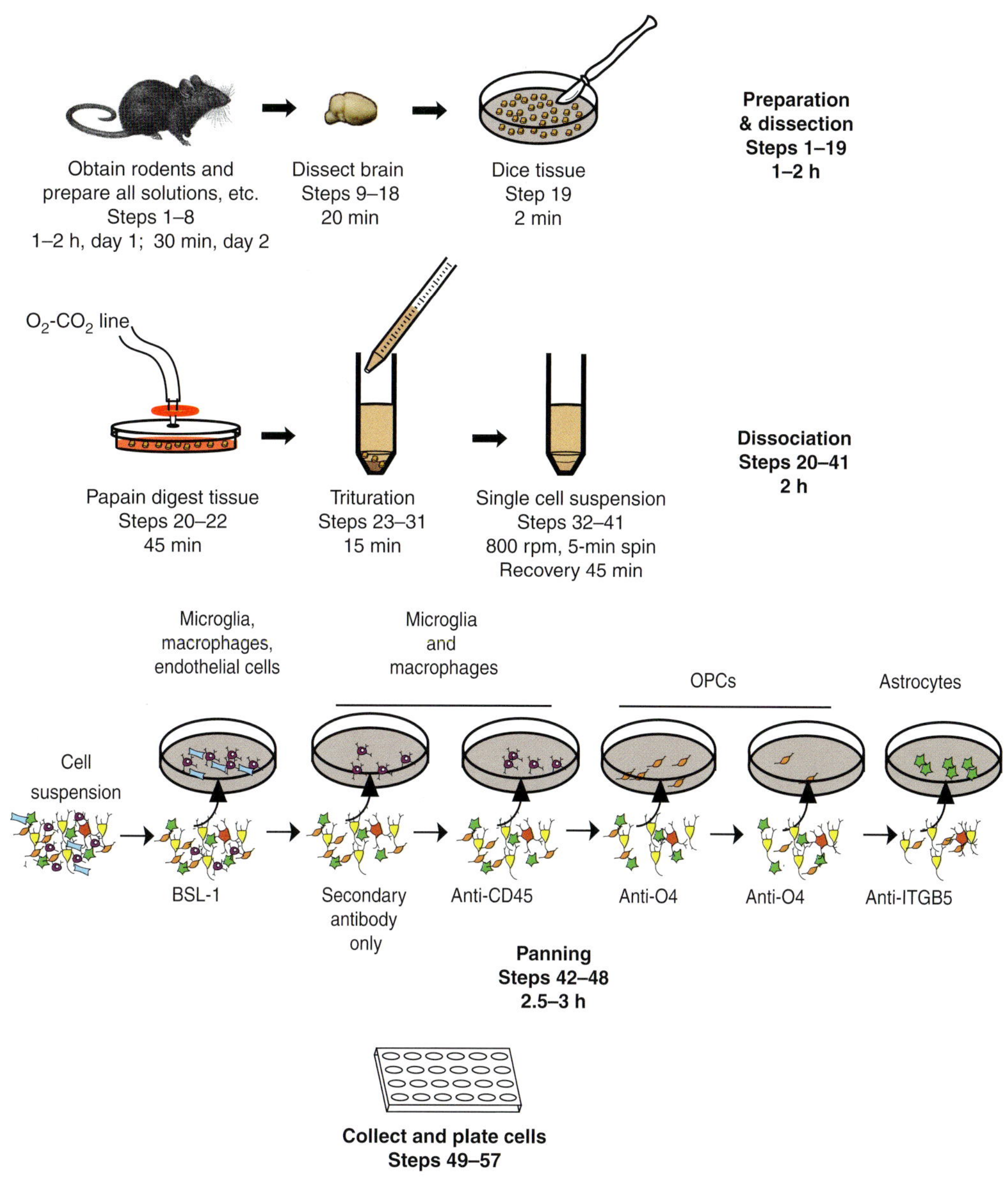

FIGURE 1. Immunopanning of astrocytes with anti-ITGB5.

Plates are initially hydrophobic, but after coating overnight, plates become visibly hydrophilic. If plates are needed immediately, a quick but less preferable way to make plates is to coat with secondary antibodies for 2 h at 37°C.

Incubate the dishes with the antibody solutions overnight at 4°C.

Preparation of Solutions and Panning Dishes

2. Aliquot and equilibrate the enzyme stock solution.
   i. Aliquot 22 mL of enzyme stock solution into a 50-mL conical tube.
   ii. Break a 2-mL pipette, attach a 0.22-µm filter, and attach it to the 5% CO_2/95% O_2 line. Bubble 5% CO_2/95% O_2 through the solution until it turns from red to orange (see online Movie 1 at cshprotocols.cshlp.org).
   iii. Put the tube of equilibrated enzyme stock solution into a 34°C water bath.

   Astrocytes are exquisitely sensitive to pH changes. Yield and viability are highly affected by improper equilibration of the enzyme and dissociation media. All media in contact with cells must be properly equilibrated before use, including the growth medium that will be added.

3. Aliquot and equilibrate the inhibitor stock solution.
   i. Prepare two tubes with 21 mL per tube, and one tube with 10 mL.
   ii. Bubble CO_2 through the solution until it turns from red to orange.

      The 2-mL pipette from Step 2 can be re-used for this step. Do not put the equilibrated inhibitor stock into a water bath.
4. Prepare 60 mL of 0.2% BSA/DNase: Combine 57 mL of D-PBS with 3 mL of 4% BSA and 24 µL of 0.4% DNase.
5. Prepare 50 mL of 0.02% BSA/DNase: Combine 45 mL of D-PBS with 5 mL of 0.2% BSA/DNase and 50 µL of 0.4% DNase.
6. Prepare high- and low-ovo inhibitor solutions.
   i. Combine the following for high-ovo inhibitor solution: 10 mL of inhibitor stock solution + 2 mL of 10× high-ovomucoid stock solution + 20 µL of 0.4% DNase.
   ii. Combine the following for low-ovo inhibitor solution (prepare two conical tubes): 21 mL of inhibitor stock solution + 1.5 mL of 10× low-ovomucoid stock solution + 100 µL of 0.4% DNase.
7. Finish preparing the panning dishes from Step 1.
   i. Wash each dish except for the BSL-1 dish three times with D-PBS.
   ii. Coat with the appropriate primary antibody solutions:
      - One "Secondary antibody only" dish: 12 mL of 0.2% BSA/DNase
      - One "BSL-1" dish: None (leave unwashed and uncoated)
      - One "CD45" dish: 20 µL rat anti-mouse CD45 in 12 mL of 0.2% BSA/DNase
      - Two "O4" dishes: 4 mL O4 hybridoma supernatant mouse IgM in 8 mL of 0.2% BSA/DNase
      - One "ITGB5" dish: 20 µL of mouse anti-human ITGB5 in 12 mL of 0.2% BSA/DNase

      Allow plates to incubate with primary antibody for at least 2 h at room temperature before use in Step 42.
8. Add 100 units of papain to the 22 mL of equilibrated enzyme stock solution from Step 2. Add 0.0036–0.0042 g of L-cysteine. Warm the mixture in a 34°C water bath before beginning the dissection.

   To maximize the activity of the enzyme, do not add it to the enzyme stock solution if it is not going to be used within 20 min. Preactivation of papain before usage optimizes the enzymatic activity, enabling shorter enzymatic dissociation times.

Brain Dissection

Dissect up to 8 P7 rat brains or 10 P7 mouse brains. Dissection should only take ~20 min (see online Movie 2 at cshprotocols.cshlp.org).

9. Prepare two 6-cm Petri dishes with 10 mL of D-PBS per dish.
10. Grab the animal to be dissected by the scruff and swiftly decapitate it with scissors. Discard the body.
11. Insert medium-sized scissors under the skin of the animal's head, avoiding the skull. Slide the scissors down the middle of the head to slice the skin into two halves. Gently pull apart the skin to reveal the skull.
12. Insert curved scissors into the base of the spinal cord, and make two cuts on each side. On one side, cut around the brain and horizontally across the part of the skull that covers the olfactory bulb.
13. Remove the skull flap, and cut the olfactory bulb with curved scissors.

Cite this protocol as *Cold Spring Harb Protoc*; doi:10.1101/pdb.prot074211

14. Gently insert curved forceps under the brain. Cut the optic nerve and transfer the brain into one of the D-PBS-containing Petri dishes from Step 9. As more animals are dissected, split the brains evenly between the two dishes.

15. Under a dissecting microscope, remove the midbrain, hindbrain, and striatal regions, leaving only the cortex.

16. Remove the meninges from the surface of the cortex by peeling them off with forceps.

    This is easier to do with younger animals but can be done until P10 on rats and ~P7 on mouse. Be careful not to make little tears in the surface of the cortex, so the meninges can be peeled off in sheets rather than in little sections.

17. Add a 200-µL drop of D-PBS to the center of two 6-cm Petri dishes.

18. Use a perforated scoop to transfer the cortices to the drops of D-PBS. Split the cortices evenly between the dishes. Put a maximum of four rat cortices or five mouse cortices per Petri dish, to enable more thorough enzymatic dissociation.

19. Use a No. 10 scalpel blade to dice the brains into ~1 mm^3 pieces (see online Movie 3 at cshprotocols.cshlp.org).

Dissociation of the Cells

After papain treatment to loosen contacts in the extracellular matrix, the tissue is washed and then mechanically dissociated by gentle sequential trituration using a 5-mL pipette with fresh inhibitor solution to yield a suspension of single cells. This dissociation needs to be quick yet gentle to tease the cells apart rather than rip them apart; ripping will result in poor cell health.

20. In a sterile hood, attach a 0.22-µm filter to a 20-mL syringe. Filter and discard 2 mL of the enzyme stock solution prepared in Step 8. Then filter 10 mL of enzyme stock solution into each of the two Petri dishes containing finely diced brains (see online Movie 3 at cshprotocols.cshlp.org).

21. Add 50 µL of 0.4% DNase to each Petri dish and swirl the dishes.

22. Allow the brains to digest by leaving the dishes for 40 min on a heat block set to 34°C. Cover with lids that have holes in the top to accommodate a 0.22-µm filter attached to the 5% CO_2/95% O_2 line. Bubble CO_2 over the brains continuously and shake the dishes every 10–15 min (see online Movie 3 at cshprotocols.cshlp.org).

    It is important to shake the dishes periodically!

23. Equilibrate 20 mL of 30% FCS and 8 mL of EBSS in a 37°C, 10% CO_2 incubator.

24. Transfer the digested cortices from both Petri dishes (Step 22) into a 50-mL conical tube. Wait for the tissue to settle, and then aspirate and discard the excess liquid with a suction pump.

25. Wash the cells by adding 4.5 mL of low-ovo inhibitor solution from one of the tubes ("Tube 1") prepared in Step 6.ii. Wait for the cells to settle, then aspirate and discard the excess liquid.

26. Repeat Step 25 four times.

    There should be ~4 mL of low-ovo inhibitor solution remaining in Tube 1. This is the low-ovo inhibitor solution to which single cells will be added in Step 30.

27. Add 4 mL of low-ovo inhibitor solution from the second tube prepared in Step 6.ii ("Tube 2") to the conical tube of cells for trituration.

28. Use a 5-mL serological pipette to quickly suck up and release the solution of brain + low-ovo inhibitor solution (see online Movie 4 at cshprotocols.cshlp.org). As the tissue dissociates, the low-ovo inhibitor solution in the tube will become cloudy. Be careful not to introduce bubbles. To minimize the introduction of CO_2 into the solution, do not lift the 5-mL pipette out of the solution.

    The dissociation buffer (the low-ovo inhibitor solution) contains Earle's balanced salts, a bicarbonate-based buffer that requires careful equilibration with 5% CO_2/95% O_2 gas before use and during papain treatment. When the dissociation buffer is exposed to room air during trituration, minimizing surface area and avoiding *bubbles is essential for maintaining the proper pH and cell health. When first using this procedure, poor cell health and viability is common, because it is hard to work quickly and efficiently. As the user becomes more practiced, an increase in cell health should be seen.*

29. Allow the brain chunks to settle.
30. Collect the single cells (the cloudy solution on top of the brain chunks) with a 1-mL pipette. Add them to the 4 mL of low-ovo inhibitor solution in Tube 1 from Step 26. Avoid transferring the chunks of brain.
31. Repeat the trituration (Steps 27–30) until 95% of the brain chunks are gone.

    Near the end of the procedure, it is normal to have pieces of tissue that cannot be dissociated with gentle pipetting. These pieces of tissue should be discarded, because harsher trituration will results in the cells being ripped apart and poor cell health. (This does not appear to be the case in other procedures, such as oligodendrocyte precursor cell [OPC] preparation, where the cells can tolerate hard pipetting.)
32. Count the cells.
    i. Dilute the cells 1:5 by combining 20 µL of cells with 80 µL of 0.02% BSA/DNase.
    ii. Dilute this mixture 1:2 with trypan blue and count the cells using a hemacytometer.

        Expect approximately 12 million cells per P7 rat brain.
33. Use a 2-mL pipette to remove brain chunks that have settled at bottom of the conical tube with the single cell suspension.
34. Use a 10-mL pipette to carefully layer 12 mL of high-ovo inhibitor solution under the single cell suspension (see online Movie 4 at cshprotocols.cshlp.org).

    After this step, there should be a clear layer of liquid beneath a cloudy cell suspension.
35. Centrifuge the cells at 110 rcf for 5 min in a tabletop centrifuge at room temperature.

    The idea here is for the dissociated cells to move down through the high-ovo solution to ensure the complete inhibition of the papain.

    Not all of the cells will come down at 110 rcf. Astrocytes at this age tend to be relatively fragile, and viability is decreased if this centrifugation step occurs at a higher speed.
36. Aspirate and discard the supernatant. There should be a visible pellet of cells at the bottom of the conical tube.
37. Resuspend the cell pellet.
    i. Add 3 mL of 0.02% BSA/DNase, and gently resuspend the cells by pipetting up and down with a 1-mL pipette.
    ii. After the cells are resuspended, add enough 0.02% BSA/DNase to bring the volume up to 9 mL.
38. Make a Nitex filter cone (see online Movie 5 at cshprotocols.cshlp.org), and pre-wet the filter with 1 mL of 0.02% BSA/DNase. Use flamed forceps (spray with ethanol and use a Bunsen flame to sterilize) to hold Nitex filter in place.
39. Filter the resuspended cells through the Nitex mesh to eliminate remaining clumps of cells and chunks of tissue.
    i. Filter 1 mL of resuspended cells at a time.
    ii. After the cells have passed through the filter, wash it with 3 mL of 0.02% BSA/DNase.
40. Count the cells.
    i. Dilute the cells 1:5 by combining 20 µL of cells with 80 µL of 0.02% BSA/DNase.
    ii. Dilute this mixture 1:2 with trypan blue and count the cells using a hemacytometer.

        There should be ~8–10 million cells per P7 pup at this stage.
41. Allow the cells to recover by incubating them at 37°C for 30–45 min in a 10% CO_2 incubator.

    This step is critical, because it allows antigens such as ITGB5 to return to the cell surface after papain digestion.

Cite this protocol as *Cold Spring Harb Protoc*; doi:10.1101/pdb.prot074211

Panning

All panning steps should be done on a flat surface at room temperature.

42. Wash the panning dishes from Step 7 three times with D-PBS (see online Movie 5 at cshprotocols.cshlp.org).
43. Add the cells to the "secondary only" dish. Swirl the dish to distribute the cells evenly. Let the dish sit for 5 min, shake, and then leave to rest for another 5 min (10 min total).

    Cell health is determined by morphology and viability by trypan blue exclusion. We have found that examining cell morphology on the first panning plate is often the best indicator of a gentle, successful procedure. A lack of floating debris indicates good health and dissociation.

44. Transfer the unbound cells (the supernatant) to the "BSL-1" dish. Swirl the dish to distribute the cells evenly. Let the dish sit for 5 min, shake, and then leave to rest for another 5 min (10 min total).

    BSL-1 pulls down a few astrocytes, so it is critical that the cells are not left for longer than 10 min.

45. Transfer the unbound cells to the "CD45" dish. Swirl the dish to distribute the cells evenly. Let the dish sit for 10 min, shake, and then leave to rest for another 10 min (20 min total).
46. Transfer the unbound cells to one of the "O4" dishes. Swirl the dish to distribute the cells evenly. Let the dish sit for 7.5 min, shake, and then leave to rest for another 7.5 min (15 min total). Repeat with the second "O4" dish.
47. Transfer the unbound cells to the "ITGB5" dish. Swirl the dish to distribute the cells evenly. Let the dish sit for 20 min, shake, and then leave to rest for another 20 min (40 min total).
48. Wash the ITGB5 dish approximately eight times with D-PBS.

    This dish has the astrocytes bound to it.

49. Add 200 units of trypsin to 8 mL of equilibrated EBSS. Add this mixture to the washed ITGB5 dish, and incubate the dish in a 37°C/10% CO_2 incubator for 3 min. Remove the dish from the incubator and tap its side.
    i. If cells do not come off easily by tapping the dish, return them to the incubator.
    ii. When cells do come off easily, continue to dislodge them by using a 10-mL serological pipette to systematically squirt them with 10 mL of 30% FCS (see online Movie 6 at cshprotocols.cshlp.org).

    Some contaminating cells (microglia, macrophages) will remain stuck to the plate. These cells are blue under the phase microscope; leave them stuck to the plate to ensure purity of the astrocytes.

    Movie 6 depicts pericyte trypsinization, which is done with two 10-cm plates. The concept and procedure is exactly the same for the astrocyte prep, except for the use of only one 15-cm plate and a 10-mL serological pipette to dislodge the cells. Do not leave trypsin at room temperature for an extended period of time. Inactivating trypsin leads to the user shearing the cells off the panning dish rather than lifting them off whole and will thus lead to drastically lower yield.

50. Transfer the dislodged cells to a 50-mL conical tube.
51. Squirt another 10 mL of 30% FCS on the dish and then transfer to the conical tube.
52. Dilute the cells by half in trypan blue and count them.

    The yield should be ~1 million cells per P7 rat pup.

53. Add 100 µL of 0.4% DNase per 10 mL of suspended cells. No incubation is needed at this point.
54. Centrifuge the cells at 170 rcf for 11 min in a tabletop centrifuge at room temperature, and then aspirate and discard the supernatant.
55. Resuspend the cell pellet in 0.02% BSA/DNase or IP-astrocyte base medium.
56. Preplate cells in 50 µL of IP-astrocyte base medium at 2,500 cells/well in a 24-well plate containing PDL-coated ACLAR plastic coverslips. Incubate the cells in a 37°C/10% CO_2 incubator for 30 min to 1 h.

    Alternatively, cells can be added directly to preequilibrated HBEGF-containing medium in a 15-cm tissue culture plate. Do not plate too many astrocytes into a dish, or they will starve themselves and overgrow. Typically, we plate 200,000–500,000 cells per 10-cm dish and 1–2 million cells per 15-cm dish; we grow them a maximum of 2 wk when plated at this density. Astrocytes will divide over time; thus the user needs to determine how many cells to plate based on how long after plating the cells will be used.

57. Carefully add 500 µL of IP-astrocyte base medium containing 5 ng/mL HBEGF to each well of the 24-well plate. Perform a half medium change every 7 d with fresh 5 ng/mL HBEGF.

Modifications for Mouse Immunopanning

The procedure for immunopanning for mouse astrocytes is basically the same as that for rat astrocytes, with a few modifications. Instead of using six immunopanning plates, only five are used. No BSL-1 dish is used, but an L1 plate to select for neurons is used before the ITGB5 plate. For the ITGB5 plate, a choice of two secondary antibodies can be used, depending on which primary antibody the user selects.

Mouse Immunopanning Plates

The following list of plates should be substituted for those described in Step 1 of the rat protocol:

- One "Secondary antibody only" dish: 60 µL goat anti-rat IgG (H + L) (Jackson ImmunoResearch 112-005-167)
- One "CD45" dish: 60 µL goat anti-rat IgG (H + L)
- One "O4" dish: 60 µL goat anti-mouse IgM µ-chain specific
- One "L1" dish: 60 µL goat anti-mouse IgG + IgM (H + L)
- One "ITGB5" dish: 60 µL Donkey anti-sheep IgG (H + L) ML* (Jackson ImmunoResearch 713-005-147) *or* 60 µL Donkey anti-goat IgG (H + L) minimal cross-reactivity (Jackson ImmunoResearch 705-005-003), depending on which primary antibody the user selects (either sheep or goat ITGB5)

Prepare the mouse panning dishes by coating five 15-cm Petri dishes with 25 mL of 50 mM Tris-HCl (pH 9.5) per dish and the appropriate secondary antibody solution. Incubate the dishes with the antibody solutions overnight at 4°C.

The following list of primary antibody solutions should be used to treat the mouse immunopanning plates:

- One "Secondary antibody only" dish: 12 mL of 0.2% BSA/DNase
- One "CD45" dish: 20 µL rat anti-mouse CD45 in 12 mL of 0.2% BSA/DNase
- One "O4" dish: 4 mL O4 hybridoma supernatant mouse IgM in 8 mL of 0.2% BSA/DNase
- One "L1" dish: 10 µL of primary antibody (Millipore MAB 5272)
- One "ITGB5" dish (no further coating is needed at this time; anti-ITGB5 antibodies will be added directly to the cell suspension, and then cells will be transferred to this dish)

These solutions substitute for those described in Step 7 of the rat protocol. Allow the coated plates to incubate with primary antibody for at least 2 h at room temperature before use.

Mouse Immunopanning Sequence

Dissect the brains and triturate the cells as described in the rat protocol. After washing the panning dishes as in Step 42 of the rat protocol, continue with the following sequence of incubations on a flat surface at room temperature.

1. Add the cells to the "secondary only" dish. Swirl the dish to distribute the cells evenly. Let the dish sit for 10 min, shake, and then leave to rest for another 10 min (20 min total).
2. Transfer the unbound cells (the supernatant) to the "CD45" dish. Swirl the dish to distribute the cells evenly. Let the dish sit for 10 min, shake, and then leave to rest for another 10 min (20 min total).

Cite this protocol as *Cold Spring Harb Protoc*; doi:10.1101/pdb.prot074211

3. Transfer the unbound cells to the “O4” dish. Swirl the dish to distribute the cells evenly. Let the dish sit for 10 min, shake, and then leave to rest for another 10 min (20 min total).
4. Transfer the unbound cells to the “L1” dish. Swirl the dish to distribute the cells evenly. Let the dish sit for 15 min, shake, and then leave to rest for another 15 min (30 min total).
5. Transfer the unbound cells to a 50-mL conical tube.
6. Add 100 µL of 0.2 µg/µL (20 µg total) sheep anti-ITGB5 antibody (R&D Systems AF3824) to the cell suspension *or* add 100 µL of 0.2 µg/µL (20 µg total) goat anti-ITGB5 (Santa Cruz SC-5401).

   The choice of antibodies depends on which secondary antibody was used to coat the immunopanning plate.
7. Mix the antibody–cell solution on a nutator or shake gently on a rocker for 30 min.
8. Add 100 µL of 0.4% DNase per 10 mL of suspended cells. No incubation is needed at this point.
9. Centrifuge at 170 rcf in a tabletop centrifuge at room temperature for 10 min, and then aspirate and discard the supernatant.
10. Resuspend the cell pellet in 10 mL of 0.02% BSA/DNase. Transfer the cells to the donkey anti-sheep secondary dish. Let the dish sit for 15 min, shake, and then leave to rest for another 15 min (30 min total).

    Wash the dish in D-PBS and continue with trypsinization and plating of the cells as described for rats (see Step 48 of the rat protocol).

RELATED INFORMATION

The purification procedures are based on previously described dissociation (Huettner and Baughman 1986; Segal et al. 1998) and immunopanning purification protocols for other cell types (Meyer-Franke et al. 1995; Barres et al. 1988, 1992).

RECIPES

Enzyme Stock Solution

Reagent	Volume	Final concentration
EBSS (10×) (Sigma-Aldrich E7510)	20 mL	1×
D^{+}-Glucose (30%)	2.4 mL	0.46%
$NaHCO_3$ (1 M)	5.2 mL	26 mM
EDTA (50 mM)	2 mL	0.5 mM
ddH_2O	170.4 mL	

Bring the volume to 200 mL with ddH_2O and filter-sterilize through a 0.22-µm filter. Store at 4°C.

High-Ovomucoid Stock Solution (10×)

To prepare, add 6 g of BSA (Sigma-Aldrich A8806) to 150 mL D-PBS. Add 6 g of trypsin inhibitor (Worthington LS003086) and mix to dissolve. Add at least 1.5 mL of 1 N NaOH to adjust the pH; continue adding NaOH as necessary to bring up the pH to 7.4. Bring the volume to 200 mL with D-PBS. Filter-sterilize through a 0.22-µm filter. Make 1.0-mL aliquots and store at −20°C.

Inhibitor Stock Solution

Reagent	Volume	Final concentration
EBSS (10×) (Sigma-Aldrich E7510)	50 mL	1×
D$^+$-glucose (30%)	6 mL	0.46%
$NaHCO_3$ (1 M)	13 mL	26 mM
ddH_2O	170.4 mL	

Bring the volume to 500 mL with ddH_2O and filter-sterilize through a 0.22-µm filter. Store at 4°C.

IP-Astrocyte Base Medium

Reagent	Final concentration
Neurobasal Medium (Gibco/Life Technologies 21103)	50%
DMEM (Gibco/Life Technologies 11960-044)	50%
Penicillin-streptomycin (Gibco/Life Technologies 15140-122)	100 U/mL (penicillin) 100 µg/mL (streptomycin)
Sodium pyruvate (100 mM; Gibco/Life Technologies 11360-070)	1 mM
L-Glutamine (200 mM; Gibco/Life Technologies 25030-081)	292 µg/mL
SATO supplement, neurobasal-based (100×) <R>	1×
NAC stock (5 mg/mL) <R>	5 µg/mL

Filter-sterilize through a 0.22-µm filter. Store at 4°C in the dark.

Low-Ovomucoid Stock Solution (10×)

To prepare, add 3 g of BSA (Sigma-Aldrich A8806) to 150 mL D-PBS. Mix well. Add 3 g of trypsin inhibitor (Worthington LS003086) and mix to dissolve. Add ~1 mL of 1 N NaOH to adjust the pH to 7.4. Bring the volume to 200 mL with D-PBS. Filter-sterilize through a 0.22-µm filter. Make 1.0-mL aliquots and store at −20°C.

NAC Stock (5 mg/mL)

To prepare, dissolve 50 mg of *N*-acetyl-L-cysteine (NAC) powder (Sigma-Aldrich A8199) in 10 mL of Neurobasal Medium (Gibco/Life Technologies 21103). (The solution will be yellowish.) Filter through a 0.22-µm filter. Prepare 20- and 80-µL aliquots and store them frozen at −20°C.

SATO Supplement, NB-based (100×)

1. Prepare the following stock solutions (these should be made fresh; do not reuse).
   - Combine 2.5 mg of progesterone (Sigma-Aldrich P8783) and 100 µL of ethanol to make a progesterone stock solution.
   - Combine 4.0 mg of sodium selenite (Sigma-Aldrich S5261), 10 µL of 1 N NaOH, and 10 mL of Neurobasal (NB, Gibco 21103-049) to make a sodium selenite stock solution.
2. Add the following to 80 mL of Neurobasal medium:

 Cite this protocol as *Cold Spring Harb Protoc;* doi:10.1101/pdb.prot074211

Reagent	Quantity	Final concentration in medium (1×)
BSA (Sigma-Aldrich A4161)	800 mg	100 µg/mL
Transferrin (Sigma-Aldrich T1147)	800 mg	100 µg/mL
Putrescine dihydrochloride (Sigma-Aldrich P5780)	128 mg	16 µg /mL
Progesterone stock solution	20 µL	60 ng/mL (0.2 µM)
Sodium selenite stock solution	800 µL	40 ng/mL

3. Mix well, and filter-sterilize through a prerinsed 0.22-µm filter. Make 200-µL or 800-µL aliquots, and store at −20°C.

ACKNOWLEDGMENTS

We thank Dr. Jim Huettner for helpful comments on the adaptation of his neuronal dissociation procedure to allow for glial purification.

REFERENCES

Bansal R, Pfeiffer SE. 1989. Reversible inhibition of oligodendrocyte progenitor differentiation by a monoclonal antibody against surface galactolipids. *Proc Natl Acad Sci* **86:** 6181–6185.

Barres BA, Silverstein BE, Corey DP, Chun LL. 1988. Immunological, morphological, and electrophysiological variation among retinal ganglion cells purified by panning. *Neuron* **1:** 791–803.

Barres B, Hart I, Coles H, Burne J, Voyvodic J, Richardson W, Raff M. 1992. Cell death and control of cell survival in the oligodendrocyte lineage. *Cell* **70:** 31–46.

Huettner JE, Baughman RW. 1986. Primary culture of identified neurons from the visual cortex of postnatal rats. *J Neurosci* **6:** 3044–3060.

Meyer-Franke A, Kaplan M, Pfrieger F, Barres B. 1995. Characterization of the signaling interactions that promote the survival and growth of developing retinal ganglion cells in culture. *Neuron* **15:** 805–819.

Segal M, Baughman R, Jones K. 1998. Stanford WebLogin (Culturing nerve cells).

Protocol 2

Purification of Astrocytes from Transgenic Rodents by Fluorescence-Activated Cell Sorting

Lynette C. Foo[1]

Department of Neurobiology, School of Medicine, Stanford University, Stanford, California 94305

The purification of astrocytes by fluorescence-activated cell sorting (FACS) requires that an astrocyte-specific promoter drive the expression of the green fluorescent protein (GFP). Our laboratory uses FACS to acutely isolate astrocytes from young and old tissue as well as to isolate GFP-negative neurons at the end of the FACS sorting to conduct comparative unbiased, large-scale gene expression studies. Because of the relatively harsh nature of FACS sorting, few astrocytes or neurons survive long enough after the sort to be cultured.

MATERIALS

It is essential that you consult the appropriate Material Safety Data Sheets and your institution's Environmental Health and Safety Office for proper handling of equipment and hazardous materials used in this protocol.

RECIPES: Please see the end of this protocol for recipes indicated by <R>. Additional recipes can be found online at http://cshprotocols.cshlp.org/site/recipes.

Reagents

BSA (4%)

To prepare a stock of 4% BSA in Dulbecco's phosphate-buffered saline (D-PBS), dissolve 8 g of BSA (Sigma-Aldrich A4161) in 150 mL of D-PBS (HyClone SH30264.01) at 37°C. Adjust the pH to 7.4 with ~1 mL of 1 N NaOH. Bring the volume to 200 mL. Filter through a 0.22-µm filter. Store in 1-mL aliquots at –20°C.

DNase (0.4%)

To prepare a 0.4% stock of DNase in Earle's balanced salt solution (EBSS), add 1 mL of EBSS (Sigma-Aldrich E6267) per 12,500 units of DNase (Worthington LS002007 or Sigma-Aldrich D4527). Keep on ice. Filter-sterilize, and store in 200-µL aliquots at –20°C.

Dulbecco's phosphate-buffered saline (D-PBS) containing Mg^{2+} and Ca^{2+} (HyClone SH30264.01)

Store at room temperature.

Enzyme stock solution <R>

High-ovomucoid (high-ovo) stock solution (10×) <R>

Inhibitor stock solution <R>

L-Cysteine hydrochloride monohydrate (Sigma-Aldrich C7880)

IP-astrocyte base medium containing 5 ng/mL HBEGF (Sigma-Aldrich 4643) <R>

Equilibrate all media as in Step 2.ii before use.

[1]Correspondence: lynettefoo@gmail.com

Cite this protocol as *Cold Spring Harb Protoc*; doi:10.1101/pdb.prot074229

Low-ovomucoid (low-ovo) stock solution (10×) <R>
Papain (Worthington LS003126)
PBS (Diamedix 1000-3)
Primary antibodies:

- Anti-PDGFRα (rat IgG, monoclonal antibody CD140a clone APA5; BD Biosciences-Pharmingen)
- O4 hybridoma supernatant (mouse IgM; Bansal and Pfeiffer 1989)
- Galactocerebroside (GalC) supernatant (for use with mice older than P8)
- Myelin oligodendrocyte glycoprotein (MOG) supernatant (for use with mice older than P8)
- O1 supernatant (for use with mice older than P8)

Propidium iodide (1 mg/mL)
Secondary antibodies:

- Goat anti-mouse IgG + IgM (H + L) (Jackson ImmunoResearch 115-005-044)
- Goat anti-mouse IgM μ-chain specific (Jackson ImmunoResearch 115-005-020)
- Goat anti-rat IgG (H + L) (Jackson ImmunoResearch 112-005-167)
- Donkey anti-mouse APC (eBioscience 17-4012-82) (for use with mice older than P8)

Transgenic mice, 8–10 S100β-eGFP (P0–P30) or Aldh1L1-eGFP
Tris-HCl (50 mM, pH 9.5), sterilized
Trypan blue

Equipment

ACLAR plastic coverslips (ethanol-washed and PDL-coated) or tissue culture plates (15-cm; PDL-coated)

To coat tissue culture plastic with poly-D-lysine (PDL), dilute a 100× PDL stock solution (1 mg/mL in H_2O) to 1× in sterile H_2O, and then coat the plastic for at least 30 min. Wash three times with sterile H_2O before drying or using. In practice, we coat ethanol-washed ACLAR coverslips after adding them individually to each well of a 24-well plate (see Step 59.i). ACLAR is very hydrophobic; add 500 μL of PDL to cover each coverslip completely. If the coverslip floats up, tap it down with a pipette tip. Wash the coverslips in the wells three times with sterile H_2O before adding medium.

Aspirator
Bunsen burner and ethanol
Conical tubes (50-mL)
Dissection equipment:

- Decapitation scissors, large (ROBOZ RS-6820)
- Dissecting microscope
- Dissection scissors, curved (ROBOZ RS-5675)
- Forceps, #5
- Forceps, curved
- Scalpel with No. 10 scalpel blade
- Scissors, medium-sized (ROBOZ RS-6702)
- Scoop, perforated (Moria MC17)

Equilibration setup (a source of carbon dioxide [5% CO_2/95% O_2] with a line leading to a sterile hood)
FACS machine
Forceps
Heat block preset to 34°C in a sterile hood
Hemacytometer
Hood for sterile tissue culture
Incubator preset to 37°C and 10% CO_2
Microscope, phase contrast
Nitex mesh filter (Tetko HC3-20)

Cut into 3-inch squares, wrap in small packets of foil, and autoclave.

Petri dishes, plastic (6-cm, 15-cm)

Petri dish lids (6 cm) with a hole in the center that accommodates a 0.22-µm filter

Use flamed forceps (spray with ethanol and use a Bunsen flame to sterilize) to melt the hole into the center of the lid.

Pipettes, serological (1-mL, 2-mL, 5-mL, 10-mL)

Pipettor, powered (e.g., Pipet-Aid)

Refrigerator preset to 4°C

Syringe, 20 mL

Syringe filters (0.22-µm)

Tabletop centrifuge (with 15 mL/50 mL centrifuge tube adaptors) at room temperature

Tissue culture plates, multiwell (24-well)

Water bath preset to 34°C

METHOD

In the method described here, brain tissue is dissected from transgenic rodents, a single-cell suspension is prepared, and cells are immunopanned before proceeding to FACS (see Fig. 1). Cells from animals older than P8 undergo additional steps before FACS to further deplete oligodendrocytes and myelin. The entire protocol is a 2-d procedure; only Step 1 is performed on the first day.

Preparation of Panning Dishes

1. Prepare panning dishes by coating four 15-cm Petri dishes with 25 mL of 50 mM Tris–HCl (pH 9.5) per dish and the appropriate secondary antibody solution.

   - One "secondary antibody only" dish: 60 µL anti-mouse IgG + IgM (H + L)
   - Two "O4" dishes: 60 µL anti-mouse IgM µ-chain specific
   - One "PDGFRα" dish: 60 µL anti-rat IgG

   Incubate the dishes overnight at 4°C.

   Plates are initially hydrophobic, but after coating overnight, they become visibly hydrophilic. If plates are needed immediately, a quick but less preferable way to make plates is to coat with secondary antibodies for 2 h at 37°C.

Preparation of Solutions and Panning Dishes

2. Aliquot and equilibrate the enzyme stock solution.

   i. Aliquot 22 mL of enzyme stock solution into a 50-mL conical tube.

   ii. Break a 2-mL pipette, attach a 0.22-µm filter, and attach it to the 5% CO_2/95% O_2 line. Bubble 5% CO_2/95% O_2 through the solution until it turns from red to orange (see online Movie 1 at cshprotocols.cshlp.org).

   iii. Put the tube of equilibrated enzyme stock solution into a 34°C water bath.

   Astrocytes are exquisitely sensitive to pH changes. Yield and viability are highly affected by improper equilibration of the enzyme and dissociation media. All media in contact with cells must be properly equilibrated before use, including the growth medium that will be added.

3. Aliquot and equilibrate the inhibitor stock solution.

   i. Prepare two tubes with 21 mL per tube, and one tube with 10 mL.

   ii. Bubble CO_2 through the solution until it turns from red to orange.

   The 2-mL pipette from Step 2 can be re-used for this step. Do not put the equilibrated inhibitor stock into the water bath.

4. Prepare 60 mL of 0.2% BSA/DNase: Combine 57 mL of DPBS with 3 mL of 4% BSA and 24 µL of 0.4% DNase.

 Cite this protocol as *Cold Spring Harb Protoc*; doi:10.1101/pdb.prot074229

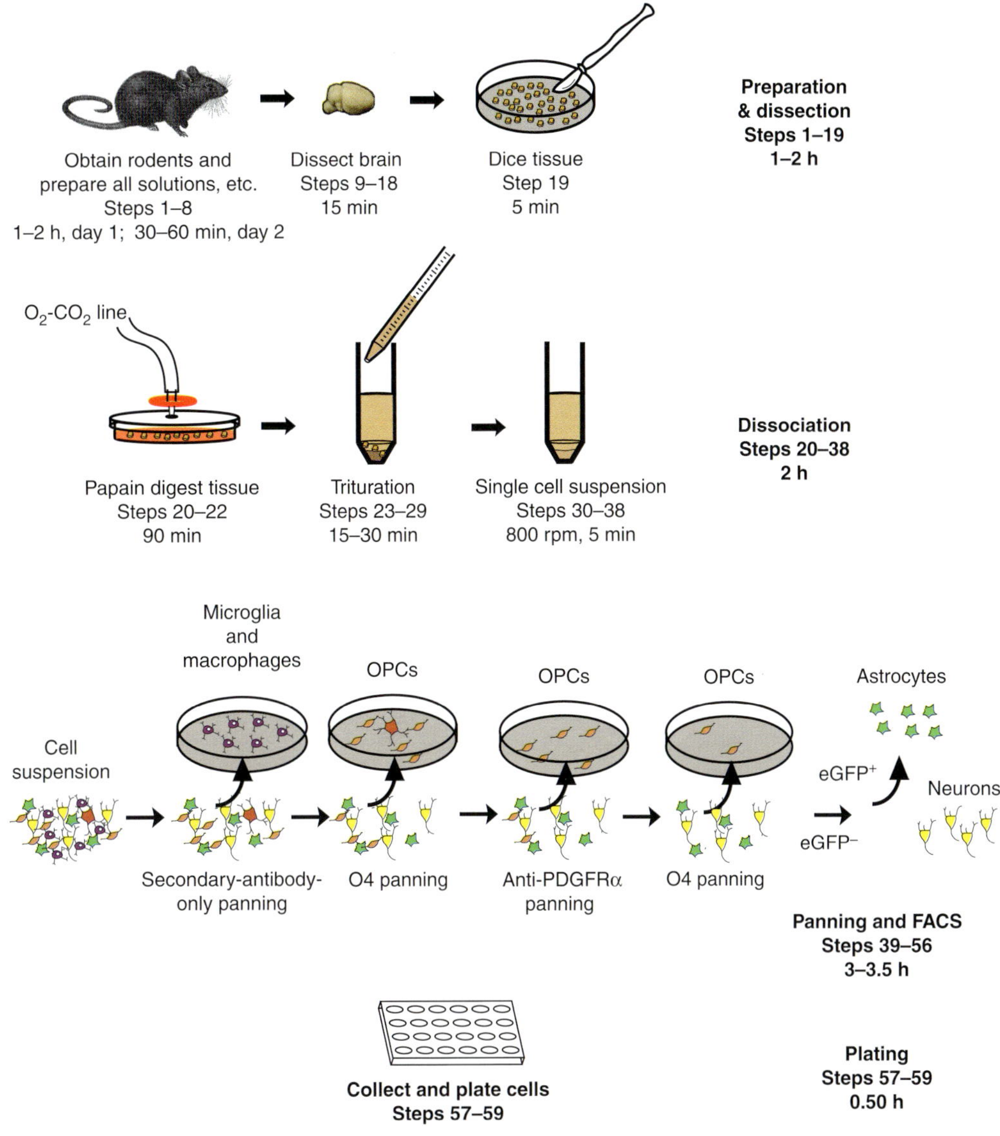

FIGURE 1. Purification of astrocytes from S100β–eGFP mice by fluorescence-activated cell sorting (FACS).

5. Prepare 50 mL of 0.02% BSA/DNase: Combine 45 mL of DPBS with 5 mL of 0.2% BSA/DNase and 50 µL of 0.4% DNase.
6. Prepare high- and low-ovo inhibitor solutions.
   i. Combine the following for the high-ovo inhibitor solution: 10 mL inhibitor stock solution (from Step 3) + 2 mL 10× high-ovomucoid stock solution + 20 µL 0.4% DNase.
   ii. Combine the following for the low-ovo inhibitor solution (prepare two conical tubes): 21 mL inhibitor stock solution (from Step 3) + 1.5 mL 10× low-ovomucoid stock solution + 100 µL 0.4% DNase.
7. Finish preparing the panning dishes from Step 1.
   i. Wash each dish three times with PBS.
   ii. Rinse each dish again with sterile PBS, and then coat them with the appropriate primary antibody solutions. Swirl the dishes for even coating, and allow them to incubate with primary antibody for at least 2 h at room temperature before use in Step 40.

- One "secondary antibody only" dish: 12 mL of 0.2% BSA/DNase
- Two "O4" dishes: 8 mL of 0.2% BSA/DNase + 4 mL of O4 hybridoma supernatant
- One "PDGFRα" dish: 12 mL of 0.2% BSA/DNase + 20 μL of PDGFRα antibody

8. After the 22 mL of enzyme stock solution from Step 2 has been warmed at 34°C for 10 min, add 100 units of papain and 3.2–4.0 mg (for 1 mM) L-cysteine-HCl. Swirl to ensure solvation. Keep the mixture warm in a 34°C water bath before beginning the dissection, but be sure to use it within 20 min of adding the papain.

   Pre-activation of papain before usage optimizes the enzymatic activity, enabling shorter enzymatic dissociation times. More papain should be used for tissue older than P7. The amount should be increased according to the age of the brain. Typically, we would use 200 units for P14 brains and up to 420 units for P30 brains. For brains older than P14, use no more than four brains per preparation to avoid overloading the panning dishes and FACS machine.

Brain Dissection

Dissection should only take ~20 min (see online Movie 2 at cshprotocols.cshlp.org).

9. Prepare two 6-cm Petri dishes with 10 mL of D-PBS per dish.
10. Grab the animal to be dissected by the scruff and swiftly decapitate it with scissors. Discard the body.
11. Insert medium-sized scissors under the skin of the animal's head, avoiding the skull. Slide the scissors down the middle of the head to slice the skin into two halves. Gently pull apart the skin to reveal the skull.
12. Insert curved scissors into the base of the spinal cord, and make two cuts on each side. On one side, cut around the brain and horizontally across the part of the skull that covers the olfactory bulb.
13. Remove the skull flap, and cut the olfactory bulb with curved scissors.
14. Gently insert curved forceps under the brain. Cut the optic nerve and transfer the brain into one of the D-PBS-containing Petri dishes from Step 9. As more animals are dissected, split the brains evenly between the two dishes. Make sure that the brains are completely covered by D-PBS.
15. Under a dissecting microscope, removing the midbrain, hindbrain, and striatal regions, leaving only the cortex.
16. Remove the meninges from the surface of the cortex by peeling them off with forceps.

    This is easier to do with younger animals but can be done until ~P7 on mouse. Be careful not to make little tears in the surface of the cortex, so the meninges can be peeled off in sheets rather than in little sections.

17. Add a 200-μL drop of D-PBS to the center of two 6-cm Petri dishes.
18. Use a perforated scoop to transfer the cortices to the drops of D-PBS. Split the cortices evenly between the dishes. Put a maximum of five mouse cortices per Petri dish, to enable more thorough enzymatic dissociation (see online Movie 3 at cshprotocols.cshlp.org).
19. Use a No.10 scalpel blade to dice the brains into ~1 mm^3 pieces.

Dissociation of the Cells

After papain treatment to loosen contacts in the extracellular matrix, the tissue is washed and then mechanically dissociated by gentle sequential trituration using a 5-mL pipette with fresh inhibitor solution to yield a suspension of single cells. This dissociation needs to be quick yet gentle to tease the cells apart rather than rip them apart; ripping will result in poor cell health.

20. In a sterile hood, attach a 0.22-μm filter to a 20-mL syringe. Filter and discard 2 mL of the papain/enzyme stock solution prepared in Step 8. Then filter 10 mL of enzyme stock solution into

 Cite this protocol as *Cold Spring Harb Protoc*; doi:10.1101/pdb.prot074229

each of the two Petri dishes containing finely diced brains (see online Movie 3 at cshprotocols.cshlp.org).

21. Add 50 µL of 0.4% DNase to each Petri dish and swirl the dishes.
22. Allow the brains to digest by leaving the dishes on a heat block set to 34°C. Cover with lids that have holes in the top to accommodate a 0.22-µm filter attached to the 5% CO_2/95% O_2 line. Blow CO_2 over the brains continuously and swirl the dishes every 10 min (see online Movie 3 at cshprotocols.cshlp.org). For P8 and younger brains, leave the dishes for 80 min. For brains older than P8, leave for 90 min.

    Avoid creating bubbles.
23. Transfer the digested cortices from both Petri dishes into a 50-mL conical tube. Wait for the tissue to settle, then aspirate and discard the excess liquid with a suction pump.
24. Wash the cells by adding 4 mL of low-ovo inhibitor solution from one of the tubes ("Tube 1") prepared in Step 6.ii. Wait for the cells to settle, and then aspirate and discard the excess liquid.
25. Repeat Step 24 three times. There should be ~5 mL of low-ovo inhibitor solution remaining in Tube 1. This is the low-ovo inhibitor solution to which single cells will be added in Step 28.
26. Add 4 mL of low-ovo inhibitor solution from the second tube prepared in Step 6.ii ("Tube 2") to the conical tube of cells for trituration.
27. Use a 5-mL serological pipette to quickly suck up and release the solution of brain + low-ovo inhibitor solution (see online Movie 4 at cshprotocols.cshlp.org). As the tissue dissociates, the low-ovo inhibitor solution in the tube will become cloudy. Be careful not to introduce bubbles, and do not lift the 5-mL pipette out of the solution.

    The dissociation buffer (the low-ovo inhibitor solution) contains Earle's balanced salts, a bicarbonate-based buffer that requires careful equilibration with 5% CO_2/95% O_2 gas before use and during papain treatment. When the dissociation buffer is exposed to room air during trituration, minimizing surface area and avoiding bubbles is essential for maintaining the proper pH and cell health. When first using this procedure, poor cell health and viability is common, because it is hard to work quickly and efficiently. As the user becomes more practiced, an increase in cell health should be seen.
28. Allow the tissue to settle. A cloudy, single-cell suspension will form above the chunks of undissociated brain. Collect the single cells with a 1-mL pipette and add them to the 5 mL of fresh low-ovo inhibitor solution left in Tube 1 from Step 25.
29. Repeat the trituration (Steps 26–28) four times, until almost all of the brain chunks have dissociated into single cells.

    Generally, 95% of the chunks are gone at the end of this step. Discard the remaining brain chunks, because their dissociation will not yield healthy cells.
30. Count the cells.
    i. Dilute the cells 1:5 by combining 20 µL of cells with 80 µL of 0.02% BSA/DNase.
    ii. Dilute this mixture 1:2 with trypan blue and count the cells using a hemacytometer.

        The cells should look healthy under the phase microscope and have a viability exceeding 90% (measured by trypan blue exclusion).
31. Use a 1-mL pipette to remove any remaining undissociated brain chunks that have settled to the bottom of the single-cell suspension.
32. Use a 10-mL pipette to carefully layer 12 mL of high-ovo inhibitor solution under the single-cell suspension (see online Movie 4 at cshprotocols.cshlp.org).

    High-ovo inhibitor solution is more dense than low-ovo inhibitor solution and will remain at the bottom, as long as its release is carefully controlled.
33. Centrifuge the cells at 110*g* for 5 min in a tabletop centrifuge at room temperature.

    The idea here is for the dissociated cells to move down through the high-ovo inhibitor solution to ensure the complete inhibition of the papain.

    Not all of the cells will come down at 110g, but viability is decreased at higher centrifugation speeds.

34. Aspirate and discard the supernatant. There should be a visible pellet of cells at the bottom of the conical tube.
35. Resuspend the cell pellet.
    i. Add 3 mL of 0.02% BSA/DNase, and gently resuspend the cells by pipetting up and down with a 1-mL pipette.
    ii. After the cells are resuspended, add 9 mL of 0.02% BSA/DNase.
36. Make a Nitex filter cone (see online Movie 5 at cshprotocols.cshlp.org), and pre-wet the filter with 1 mL of 0.02% BSA/DNase. Use flamed forceps (spray with ethanol and use a Bunsen flame to sterilize) to hold the Nitex filter in place.
37. Filter the resuspended cells through the Nitex mesh to eliminate remaining clumps of cells and chunks of tissue.
    i. Filter 1 mL of resuspended cells at a time.
    ii. After the cells have passed through the filter, wash it with 3 mL of 0.02% BSA/DNase.
38. Count the cells.
    i. Dilute the cells 1:5 by combining 20 µL of cells with 80 µL of 0.02% BSA/DNase.
    ii. Dilute this mixture 1:2 with trypan blue and count the cells using a hemacytometer.

       There should be approximately 8–10 million cells per P7 pup at this stage.

       To just isolate eGFP$^+$ astrocytes from Aldh1L1-eGFP mice by FACS, eliminate the panning steps that follow and send the dissociated single-cell suspensions directly for FACS. Otherwise, continue to panning in Step 39.

Panning

All panning steps should be done on a flat surface at room temperature.

39. Wash the first panning dish (the "secondary antibody only" dish) three times with DPBS (see online Movie 5 at cshprotocols.cshlp.org).
40. Add the filtered cells to secondary antibody panning dish. Swirl the dish to distribute the cells evenly. Let the dish sit for 7.5 min, gently swirl, and then leave to rest for another 7.5 min (15 min total).

    Cells will adhere very quickly to this plate. Neuronal morphology is a good indicator of overall health; dissociated cells from P6–P8 mice should yield suspensions in which it is common to find neurons with small-to-medium length axons and dendrites.

41. Transfer the unbound cells (the supernatant) to the first "O4" dish. Wash the "secondary antibody only" dish once with 2 mL of 0.02% BSA/DNase, and add this buffer to the "O4" dish. Swirl the dish to distribute the cells evenly. Let the dish sit for 12.5 min, gently swirl, and then leave to rest for another 12.5 min (25 min total).
42. Transfer the unbound cells to the "PDGFRα" dish. Wash the O4 dish with 2 mL of 0.02% BSA/DNase, and add this buffer to the PDGFRα dish. Swirl the dish to distribute the cells evenly. Let the dish sit for 15 min, gently swirl, and then leave to rest for another 15 min (30 min total).
43. Transfer the unbound cells to the second "O4" dish. Do not wash the PDGFRα dish, because the cells only weakly adhere to this dish. Swirl the dish to distribute the cells evenly. Let the dish sit for 10 min, gently swirl, and then leave to rest for another 10 min (20 min total).
44. Transfer the unbound cells to a 50-mL conical tube.
45. Count the cells.
    i. Dilute the cells 1:5 by combining 20 µL of cells with 80 µL of 0.02% BSA/DNase.
    ii. Dilute this mixture 1:2 with trypan blue and count the cells using a hemacytometer.

Cite this protocol as *Cold Spring Harb Protoc*; doi:10.1101/pdb.prot074229

For animals P8 and younger, proceed to Step 54 (FACS). For animals older than P8, that have begun to myelinate, perform additional depletion of oligodendrocytes and myelin in Steps 46–53 before continuing to FACS.

Additional Depletion for Animals Older than P8

Perform Steps 46–53 for additional depletion of oligodendrocytes and myelin.

46. Harvest the nonadherent cells from the second O4 dish by centrifugation at 170*g* for 5 min in a tabletop centrifuge at room temperature.
47. Resuspend the cells at 20×10^6 cells/5 mL (4×10^6 cells/mL) in 0.02% BSA/DNase containing 1:30 dilutions of galactocerebroside (GalC), myelin oligodendrocyte glycoprotein (MOG), and O1 hybridoma supernatants.
48. Incubate the cell suspension for 15 min at room temperature.
49. Wash the cells.
    i. Dilute with 40 mL of 0.02% BSA/DNase.
    ii. Centrifuge at 170*g* for 5 min in a tabletop centrifuge at room temperature. Aspirate and discard the supernatant.
50. Resuspend in 4 mL of 0.02% BSA/DNase containing 20 µg of donkey anti-mouse APC.
51. Incubate for 15 min at room temperature.
52. Wash out the secondary antibody by diluting the cells in 40 mL of 0.02% BSA/DNase.
53. Count the cells.
    i. Dilute the cells 1:5 by combining 20 µL of cells with 80 µL of 0.02% BSA/DNase.
    ii. Dilute this mixture 1:2 with trypan blue and count the cells using a hemacytometer.
    Continue to Step 54 (FACS).

FACS

54. Centrifuge the cells at 170*g* for 5 min in a tabletop centrifuge at room temperature. Aspirate and discard the supernatant.
55. Resuspend the cells to a concentration of 10×10^6 cells/mL in 0.02% BSA/DNase with 8 µL of 1 mg/mL propidium iodide per 3 mL of cells, and sort the cells by FACS, as described by Cahoy et al. (2008).
56. Sort the cells using FACS, as described by Cahoy et al. (2008).
    Each P7 pup brain should yield approximately 15×10^6 cells and ~300,000 astrocytes.

Plating the Cells

57. After FACS, centrifuge the cells at 170*g* for 10 min in a tabletop centrifuge at room temperature.
58. Resuspend the cells in 0.02% BSA/DNase.
59. Plate the cells.
    i. Carefully add 500 µl of IP-astrocyte base medium with HB-EGF to each well of a 24-well plate containing PDL-coated ACLAR plastic coverslips.
    Alternatively, PDL-coated tissue culture dishes can be used without coverslips.
    ii. Plate the cells at 10,000 cells/well.
    iii. Centrifuge the plated cells at 40*g* for 3 min in a tabletop centrifuge at room temperature.
    iv. Incubate the cells at 37°C in a 10% CO_2 incubator.

RELATED INFORMATION

The purification procedures are based on previously described dissociation (Huettner and Baughman 1986; Segal et al. 1998) and immunopanning purification protocols for other cell types (Barres et al. 1992, 1988; Meyer-Franke et al. 1995).

RECIPES

Enzyme Stock Solution

Reagent	Volume	Final concentration
10× EBSS (Sigma-Aldrich E7510)	20 mL	1×
30% D$^+$-glucose	2.4 mL	0.46%
1 M $NaHCO_3$	5.2 mL	26 mM
50 mM EDTA	2 mL	0.5 mM
ddH_2O	170.4 mL	

Bring the volume to 200 mL with ddH_2O and filter-sterilize through a 0.22-µm filter.

High-Ovomucoid Stock Solution (10×)

To prepare, add 6 g of BSA (Sigma-Aldrich A8806) to 150 mL D-PBS. Add 6 g of trypsin inhibitor (Worthington LS003086) and mix to dissolve. Add at least 1.5 mL of 1 N NaOH to adjust the pH; continue adding NaOH as necessary to bring up the pH to 7.4. Bring the volume to 200 mL with D-PBS. Filter-sterilize through a 0.22-µm filter. Make 1.0-mL aliquots, and store at −20°C.

Inhibitor Stock Solution

Reagent	Volume	Final concentration
10× EBSS (Sigma-Aldrich E7510)	50 mL	1×
30% D$^+$-glucose	6 mL	0.46%
1 M $NaHCO_3$	13 mL	26 mM
ddH_2O	170.4 mL	

Bring the volume to 500 mL with ddH_2O and filter-sterilize through a 0.22-µm filter.

IP-Astrocyte Base Medium

Reagent	Final concentration
Neurobasal Medium (Gibco/Life Technologies 21103)	50%
DMEM (Gibco/Life Technologies 11960-044)	50%
Penicillin-streptomycin (Gibco/Life Technologies 15140-122)	100 U/mL (penicillin) 100 µg/mL (streptomycin)
Sodium pyruvate (100 mM; Gibco/Life Technologies 11360-070)	1 mM
L-glutamine (200 mM; Gibco/Life Technologies 25030-081)	292 µg/mL
SATO supplement, neurobasal-based (100×) <R>	1×
NAC stock (5 mg/mL) <R>	5 µg/mL

Filter-sterilize through a 0.22-µm filter. Store at 4°C.

Cite this protocol as *Cold Spring Harb Protoc*; doi:10.1101/pdb.prot074229

Low-Ovomucoid Stock Solution (10×)

To prepare, add 3 g of BSA (Sigma-Aldrich A8806) to 150 mL D-PBS. Mix well. Add 3 g of trypsin inhibitor (Worthington LS003086) and mix to dissolve. Add ~1 mL of 1 N NaOH to adjust the pH to 7.4. Bring the volume to 200 mL with D-PBS. Filter-sterilize through a 0.22-µm filter. Make 1.0-mL aliquots and store at −20°C.

ACKNOWLEDGMENTS

We thank Dr. Jim Huettner for helpful comments on the adaptation of his neuronal dissociation procedure to allow for glial purification.

REFERENCES

Bansal R, Pfeiffer SE. 1989. Reversible inhibition of oligodendrocyte progenitor differentiation by a monoclonal antibody against surface galactolipids. *Proc Natl Acad Sci* **86:** 6181–8185.

Barres BA, Silverstein BE, Corey DP, Chun LL. 1988. Immunological, morphological, and electrophysiological variation among retinal ganglion cells purified by panning. *Neuron* **1:** 791–803.

Barres B, Hart I, Coles H, Burne J, Voyvodic J, Richardson W, Raff M. 1992. Cell death and control of cell survival in the oligodendrocyte lineage. *Cell* **70:** 31–46.

Cahoy JD, Emery B, Kaushal A, Foo LC, Zamanian JL, Christopherson KS, Xing Y, Lubischer JL, Krieg PA, Krupenko SA, et al. 2008. A transcriptome database for astrocytes, neurons, and oligodendrocytes: A new resource for understanding brain development and function. *J Neurosci* **28:** 264–278.

Hildebrand B, Olenik C, Meyer DK. 1997. Neurons are generated in confluent astroglial cultures of rat neonatal neocortex. *Neuroscience* **78:** 957–966.

Huettner JE, Baughman RW. 1986. Primary culture of identified neurons from the visual cortex of postnatal rats. *J Neurosci* **6:** 3044–3060.

Meyer-Franke A, Kaplan M, Pfrieger F, Barres B. 1995. Characterization of the signaling interactions that promote the survival and growth of developing retinal ganglion cells in culture. *Neuron* **15:** 805–819.

Segal M, Baughman R, Jones K. 1998. Stanford WebLogin (Culturing nerve cells).

CHAPTER 6

Purification and Culture of Central Nervous System Pericytes

Lu Zhou,[1] Fabien Sohet,[2] and Richard Daneman[2,3]

[1]*Department of Neurobiology, Stanford University School of Medicine, Stanford, California 94305-5125;*
[2]*Department of Anatomy, UCSF, San Francisco, California 94143-0452*

Pericytes are found on the abluminal surface of endothelial tubes. The function of these cells is still being elucidated, but they have been shown to be important for development and maintenance of the blood–brain barrier, regulation of angiogenesis and capillary blood flow, and regulation of the neural response to injury and disease. Previously used methods to isolate pericytes have relied on negative selection and prolonged culture of microvessel cells and may lead to populations of pericytes contaminated by other neural cell types. We have developed an immunopanning protocol to specifically purify pericytes from capillaries in the rodent optic nerve. This method relies on a combination of negative and positive selection criteria and allows prospective, acute isolation of pericytes. Use of this method will facilitate studies of pericyte cell biology and function and pericyte–endothelial cell interactions.

PERICYTE MORPHOLOGY

Pericytes are a type of mural cell that adhere to the abluminal surface of endothelial tubes. On larger vessels, including the arteries, arterioles, and veins, the mural cells are vascular smooth muscle cells (vSMCs). These cells surround the abluminal surface of large vessels. In arteries, there are several layers of vSMCs that form thick walls around the endothelium; these layers are important for vessel structure. The vSMCs are characterized by the expression of smooth muscle actin and can contract in response to stimuli in order to control blood flow. At the capillary level, the mural cells are pericytes. Pericytes adhere to the abluminal surface of the capillary endothelium, forming an incomplete layer, and are embedded within the vascular basement membrane (Sims 1986). The pericytes send out finger-like processes that often reach the length of several endothelial cells and traverse branch points in the vessels (Sims 1986). The processes adhere to the endothelial cells at individual points via peg-and-socket junctions that are mediated through N-cadherin-based protein interactions (Gerhardt et al. 2000; Gerhardt and Betsholtz 2003). The abundance of pericytes on capillaries varies in different tissues. Central nervous system (CNS) capillaries have the greatest pericyte coverage, with an estimated endothelial cell-to-pericyte ratio between 3:1 and 1:1 (Sims 1986). Chick–quail chimera experiments have shown that CNS pericytes are derived from the neural crest, whereas endothelial cells are derived from the mesoderm (Bergwerff et al. 1998; Etchevers et al. 2001; Korn et al. 2002).

PERICYTE FUNCTION

Studies have demonstrated that platelet-derived growth factor-BB (PDGF-BB) secreted from angiogenic endothelial cells binds to platelet-derived growth factor receptor β (PDGFRβ) tyrosine kinase on

[3]Correspondence: richard.daneman@ucsf.edu

Cite this introduction as *Cold Spring Harb Protoc*; doi:10.1101/pdb.top070888

mural cells and is required for recruitment of pericytes to the angiogenic sprouts (Lindahl et al. 1997; Tallquist et al. 2003; Gaengel et al. 2009). Genetic mouse knockout models for PDGFB (which forms the homodimer PDGF-BB) or PDGFRβ die at birth from insufficient mural cell recruitment to the vessels; CNS blood vessels in these knockouts completely lack pericytes and often form microaneurysms that can hemorrhage (Leveen et al. 1994; Soriano 1994; Lindahl et al. 1997; Hellstrom et al. 1999). Several groups have developed mouse models with attenuated PDGFB signaling, including Tallquist et al. (2003), who developed an allelic series of PDGFRβ hypomorphs, and Lindblom et al. (2003), who developed mice whose PDGFB gene contained a deletion of the extracellular matrix retention signal. Although these mice have decreased pericyte coverage of the endothelium, they do live and have been used to study the role of pericytes in adults. Several groups have demonstrated that pericytes are required for blood–brain barrier (BBB) formation during development and maintenance in adulthood, as well as during aging (Armulik et al. 2010; Bell et al. 2010; Daneman et al. 2010). These studies showed that pericytes are required for the function of the BBB, and that the total number of pericytes is important for the relative permeability of the vessels. Pericytes appear to regulate the BBB not by inducing BBB-specific gene expression, but by inhibition of genes that increase the permeability of the vessels (Daneman et al. 2010). In particular, pericytes inhibit endothelial transcytosis and the expression of leukocyte adhesion molecules (Armulik et al. 2010; Daneman et al. 2010). Pericytes have also been shown to regulate vascular branching, vascular remodeling, and vessel diameter (Armulik et al. 2011), and evidence suggests that pericyte contraction might be important for regulating blood flow in capillaries (Peppiatt et al. 2006; Bell et al. 2010). In brain slices and retina preparations, local pericyte contraction can be elicited by electrical stimulation of pericytes or perfusion with ATP or noradrenalin (Peppiatt et al. 2006).

Dore-Duffy and colleagues have suggested that there may be pericyte heterogeneity, with multiple subclasses that can be defined based on morphology and the ability to migrate away from the vessel after neural injury (Dore-Duffy et al. 2000; Dore-Duffy 2008; Bonkowski et al. 2011). Goritz et al. (2011) further identified a specific subset of pericytes, termed Type A pericytes, that are genetically labeled in mice with a GLAST-creER transgenic line. These Type A pericytes migrate away from the vessels after spinal cord injury and are required for scar formation around the injury. It is becoming clear that pericytes are very important for regulating the neural response to injury and disease. Perhaps the most obvious example is in diabetic retinopathy, where loss of retinal pericytes is observed at the disease's earliest stage and is thought to be critical for its pathology (Hammes et al. 2002).

PURIFICATION OF PERICYTES

Dore-Duffy and colleagues have suggested that CNS pericytes may also be pluripotent neural stem cells. In vitro pericytes are capable of generating self-renewing spheres in response to basic fibroblast growth factor; these spheres can differentiate into various neural lineage cell types (Dore-Duffy et al. 2006). To purify the pericytes, this group used centrifugation to isolate microvessels and negative purification by FACS to deplete endothelial cells (Dore-Duffy 2003). They used expression of Ng2 and nestin as evidence that the purified cell population was pericytes. Both markers, however, are also expressed by other neural cell types, and because this group did not use positive selection criteria to specifically isolate pericytes, it is difficult to say that they cultured a single population of cells. Several groups have used prolonged culture of microvessels in selective conditions to isolate rat CNS pericytes that outgrew the CNS endothelial cells (Hayashi et al. 2004; Dohgu et al. 2005). These methods are also limited, in that they do not use prospective identification of pericytes to isolate a pure population, and thus may in fact contain a mixture of perivascular cells including pericytes, macrophages, and stem cells.

Our studies employ immunopanning for purification of CNS pericytes from optic nerve tissue. This method, as described in Protocol 1: Purification of Pericytes from Rodent Optic Nerve by Immunopanning (Zhou et al.), involves a combination of negative and positive selection and relies on pericytes' expression of PDGFRβ. Unlike previously used methods, immunopanning allows

Cite this introduction as *Cold Spring Harb Protoc*; doi:10.1101/pdb.top070888

prospective, acute isolation of pericytes and leads to preparations that are reliably free of contaminating cells. Use of these immunopanned cell preparations will improve our understanding of pericyte cell biology and function and pericyte–endothelial cell interactions.

REFERENCES

Armulik A, Genové G, Mäe M, Nisancioglu MH, Wallgard E, Niaudet C, He L, Norlin J, Lindblom P, Strittmatter K, et al. 2010. Pericytes regulate the blood–brain barrier. *Nature* **468:** 557–561.

Armulik A, Genové G, Betsholtz C. 2011. Pericytes: developmental, physiological, and pathological perspectives, problems, and promises. *Dev Cell* **21:** 193–215.

Bell RD, Winkler EA, Sagare AP, Singh I, LaRue B, Deane R, Zlokovic BV. 2010. Pericytes control key neurovascular functions and neuronal phenotype in the adult brain and during brain aging. *Neuron* **68:** 409–427.

Bergwerff M, Verberne ME, DeRuiter MC, Poelmann RE, Gittenberger-de Groot AC. 1998. Neural crest cell contribution to the developing circulatory system: Implications for vascular morphology? *Circ Res* **82:** 221–231.

Bonkowski D, Katyshev V, Balabanov RD, Borisov A, Dore-Duffy P. 2011. The CNS microvascular pericyte: pericyte–astrocyte crosstalk in the regulation of tissue survival. *Fluids Barriers CNS* **8:** 8.

Daneman R, Zhou L, Kebede AA, Barres BA. 2010. Pericytes are required for blood–brain barrier integrity during embryogenesis. *Nature* **468:** 562–566.

Dohgu S, Takata F, Yamauchi A, Nakagawa S, Egawa T, Naito M, Tsuruo T, Sawada Y, Niwa M, Kataoka Y. 2005. Brain pericytes contribute to the induction and up-regulation of blood–brain barrier functions through transforming growth factor-β production. *Brain Res* **1038:** 208–215.

Dore-Duffy P. 2003. Isolation and characterization of cerebral microvascular pericytes. *Methods Mol Med* **89:** 375–382.

Dore-Duffy P. 2008. Pericytes: Pluripotent cells of the blood brain barrier. *Curr Pharm Des* **14:** 1581–1593.

Dore-Duffy P, Owen C, Balabanov R, Murphy S, Beaumont T, Rafols JA. 2000. Pericyte migration from the vascular wall in response to traumatic brain injury. *Microvasc Res* **60:** 55–69.

Dore-Duffy P, Katychev A, Wang X, Van Buren E. 2006. CNS microvascular pericytes exhibit multipotential stem cell activity. *J Cereb Blood Flow Metab* **26:** 613–624.

Etchevers HC, Vincent C, Le Douarin NM, Couly GF. 2001. The cephalic neural crest provides pericytes and smooth muscle cells to all blood vessels of the face and forebrain. *Development* **128:** 1059–1068.

Gaengel K, Genové G, Armulik A, Betsholtz C. 2009. Endothelial–mural cell signaling in vascular development and angiogenesis. *Arterioscler Thromb Vasc Biol* **29:** 630–638.

Gerhardt H, Betsholtz C. 2003. Endothelial–pericyte interactions in angiogenesis. *Cell Tissue Res* **314:** 15–23.

Gerhardt H, Wolburg H, Redies C. 2000. N-cadherin mediates pericytic–endothelial interaction during brain angiogenesis in the chicken. *Dev Dyn* **218:** 472–479.

Göritz C, Dias DO, Tomilin N, Barbacid M, Shupliakov O, Frisén J. 2011. A pericyte origin of spinal cord scar tissue. *Science* **333:** 238–242.

Hammes HP, Lin J, Renner O, Shani M, Lundqvist A, Betsholtz C, Brownlee M, Deutsch U. 2002. Pericytes and the pathogenesis of diabetic retinopathy. *Diabetes* **51:** 3107–3112.

Hayashi K, Nakao S, Nakaoke R, Nakagawa S, Kitagawa N, Niwa M. 2004. Effects of hypoxia on endothelial/pericytic co-culture model of the blood–brain barrier. *Regul Pept* **123:** 77–83.

Hellström M, Kalén M, Lindahl P, Abramsson A, Betsholtz C. 1999. Role of PDGF-B and PDGFR-β in recruitment of vascular smooth muscle cells and pericytes during embryonic blood vessel formation in the mouse. *Development* **126:** 3047–3055.

Korn J, Christ B, Kurz H. Neuroectodermal origin of brain pericytes and vascular smooth muscle cells. *J Comp Neurol* **442:** 78–88.

Levéen P, Pekny M, Gebre-Medhin S, Swolin B, Larsson E, Betsholtz C. 1994. Mice deficient for PDGF B show renal, cardiovascular, and hematological abnormalities. *Genes Dev* **8:** 1875–1887.

Lindahl P, Johansson BR, Levéen P, Betsholtz C. 1997. Pericyte loss and microaneurysm formation in PDGF-B-deficient mice. *Science* **277:** 242–245.

Lindblom P, Gerhardt H, Liebner S, Abramsson A, Enge M, Hellstrom M, Backstrom G, Fredriksson S, Landegren U, Nystrom HC, et al. 2003. Endothelial PDGF-B retention is required for proper investment of pericytes in the microvessel wall. *Genes Dev* **17:** 1835–1840.

Peppiatt CM, Howarth C, Mobbs P, Attwell D. 2006. Bidirectional control of CNS capillary diameter by pericytes. *Nature* **443:** 700–704.

Sims DE. 1986. The pericyte—A review. *Tissue Cell* **18:** 153–174.

Soriano P. 1994. Abnormal kidney development and hematological disorders in PDGF beta-receptor mutant mice. *Genes Dev* **8:** 1888–1896.

Tallquist MD, French WJ, Soriano P. 2003. Additive effects of PDGF receptor beta signaling pathways in vascular smooth muscle cell development. *PLoS Biol* **1:** E52.

Zhou L, Sohet F, Daneman R. 2014. Purification of pericytes from rodent optic nerve by immunopanning. *Cold Spring Harb Protoc* doi: 10.1101/pdb.prot074955.

Protocol 1

Purification of Pericytes from Rodent Optic Nerve by Immunopanning

Lu Zhou,[1] Fabien Sohet,[2] and Richard Daneman[2,3]

[1]*Department of Neurobiology, Stanford University School of Medicine, Stanford, California 94305-5125;*
[2]*Department of Anatomy, UCSF, San Francisco, California 94143-0452*

This protocol describes the use of immunopanning to purify rodent pericytes from the optic nerve. Immunopanning permits the prospective isolation of pericytes from optic nerve tissue by relying on the binding of pericytes to an anti-PDGFRβ (platelet-derived growth factor receptor beta) antibody adhered to a Petri dish. The cells are viable at the end of this gentle procedure, and they can be analyzed acutely for gene expression or cultured alone or in coculture with other central nervous system (CNS) cell types, including CNS endothelial cells and CNS astrocytes. As written, this procedure is used for isolation of optic nerve pericytes from the rat. The same PDGFRβ antibodies can be used for purifying optic nerve pericytes from the mouse, but alternate negative panning antibodies must be used to ensure that astrocytes do not contaminate the preparation. This procedure can also be modified to purify pericytes from the brain. The same PDGFRβ antibody is used, but additional steps (specific dissections or negative panning) are required to ensure that other PDGFRβ-positive contaminants, including cells from the rostral migratory stream, are depleted from the cell suspension.

MATERIALS

It is essential that you consult the appropriate Material Safety Data Sheets and your institution's Environmental Health and Safety Office for proper handling of equipment and hazardous materials used in this protocol.

RECIPES: Please see the end of this protocol for recipes indicated by <R>. Additional recipes can be found online at http://cshprotocols.cshlp.org/site/recipes.

Reagents

Antibodies and reagents for immunostaining and identifying pericytes and contaminating cells (e.g., antibodies against NG2 and smooth muscle actin, DAPI; see Step 52)

Bovine serum albumin (BSA), 4%

To prepare 4% BSA in Dulbecco's phosphate-buffered saline (D-PBS), dissolve 8 g of BSA (Sigma-Aldrich A4161) in 150 mL of D-PBS (Invitrogen 14287-080). Adjust the pH to 7.4 with 1 N NaOH, and bring the volume to 200 mL. Filter through a prerinsed 0.22-µm filter, and store in 1.0-mL aliquots at −20°C.

DMEM (Gibco 11960-044)

Store at 4°C in the dark.

DNase stock solution (12,5000 U/mL)

To prepare DNase stock solution in Earle's balanced salt solution (EBSS), dissolve 12,500 U of DNase I (Worthington LS002007) in 1 mL of ice-cold EBSS. Filter-sterilize the solution through a prerinsed 0.22-µm filter on ice, and freeze 200-µL aliquots overnight at −20°C. Store at −20°C to −30°C.

[3]Correspondence: richard.daneman@ucsf.edu

Cite this protocol as *Cold Spring Harb Protoc*; doi:10.1101/pdb.prot074955

D-PBS (Invitrogen 14287-080)

To help ensure that the D-PBS is at a neutral pH, add the indicator phenol red (Sigma-Aldrich P0290) at a 1:1000 dilution from 0.5% stock.

D-PBS without Ca^{2+}/Mg^{2+} (Invitrogen 14190-144)

D-PBS without Ca^{2+}/Mg^{2+} is used ONLY for tissue dissection. Cells will not stick to the panning plates in Ca^{2+}/Mg^{2+}-free DPBS.

Earle's Buffered Saline Solution (EBSS, Invitrogen 14155-063)

To help ensure that the EBSS is at a neutral pH, add the indicator phenol red (Sigma-Aldrich P0290) at a 1:1000 dilution from 0.5% stock.

Fetal calf serum (FCS), heat-inactivated

Make 50-mL aliquots of FCS (Invitrogen 16000-036) and heat inactivate for 30 min in a water bath at 55°C. Store the aliquots at −20°C.

High-ovomucoid (high-ovo) stock solution (6×) <R>

Insulin (0.5 mg/mL) <R>

L-Cysteine (Sigma-Aldrich C-7477)

L-Glutamine (Invitrogen 25030-081)

Thaw this 100× solution at room temperature in a beaker, make 200- and 800-µL aliquots, and store them at −20°C.

Low-ovomucoid (low-ovo) stock solution (10×) <R>

NaOH (1 M) and HCl (1 M) for pH adjustments

Papain (Worthington Biomedical LS003126)

Penicillin/streptomycin (pen-strep, Invitrogen 15140-122)

Prepare 5-mL aliquots of this 100× solution and store at 4°C.

Phosphate-buffered saline (PBS)

Poly-D-lysine (PDL) stock (1 mg/mL) <R>

Primary antibodies (for isolating pericytes from rats)

- C5 hybridoma supernatant
- Goat anti-human PDGFRβ (R&D Systems AF385)
- Goat-anti mouse PDGFRβ (R&D Systems AF1042)

Rat pups, postnatal day 6–8 Sprague–Dawley

In theory, this protocol can be altered to isolate pericytes from other ages, but the dissociation procedure needs to be optimized to ensure that appropriate antigenicity remains. We typically purify cells from one to four litters (10 pups/litter).

Secondary antibodies (for isolating pericytes from rats)

- Donkey-anti-goat IgG (Jackson ImmunoResearch 705-005-147)
- Goat-anti-mouse IgG + IgM (H + L) (Jackson ImmunoResearch 115-005-044)

Sodium pyruvate (Invitrogen 11360-070)

Make 5-mL aliquots of this 100× stock, and store them at 4°C.

Tris-HCl (50 mM, pH 9.5)

Trypan blue (Invitrogen 15250-061)

Trypsin stock solution

Dissolve the trypsin (Sigma-Aldrich T9935) at 50,000 U/mL in EBSS on ice. If the trypsin does not dissolve easily, try stirring it with a magnetic stir bar or altering the pH with NaOH or HCl. When fully dissolved, store 200-µL aliquots at −80°C.

Equipment

CO_2 chamber and tank for euthanizing rats

Conical tubes (15-mL and 50-mL)

Dissection equipment

- Dissection microscope

Forceps (#5 or #55)
Large, sharp scissors for decapitation (e.g., Roboz RS-6820)
Small, curved scissors (e.g. Roboz RS-5675)
Very small dissection scissors (Very fine scissors (for optic nerve prep, e.g. Roboz RS-5602))
Scalpel with #10 blade

Ethanol-washed glass coverslips <R>
Filter sterilization system with 500-mL bottle
Hemacytometer
Incubator at 37°C, 10% CO_2
Microscope, phase contrast
Petri dish (10 cm)
Pipette tips (1 mL)
Sterile laminar flow hood for tissue culture, equipped with aspirator
Syringe filters (0.22 µm)
Syringe needles (21 and 23 gauge)
Syringes (e.g., 10 and 20 mL)
Tabletop centrifuge (with 15- and 50-mL tube adaptors) at room temperature
Tissue culture dishes (10 cm, six well, or 24 well)
Water baths, preset to 35°C and 37°C

METHOD

This method begins with dissection of the rat pups and creation of a single-cell suspension of optic nerve cells. These cells then undergo two immunopanning steps to purify the optic nerve pericytes before they are plated (see Fig. 1 for an overview). The entire protocol is a 2-d procedure; Steps 1–5 are performed on the first day. Perform all steps in a sterile tissue culture hood, with the exception of the dissection. All solutions in contact with cells must be sterile, and all steps except for the dissection should be performed in a sterile tissue culture hood.

Preparation of Panning Dishes, Culture Dishes, Coverslips, and Solutions

Day 1

1. Prepare panning dishes by coating 10-cm Petri dishes with secondary antibody solution. Swirl the dishes until their surfaces are evenly coated, and then incubate them overnight at 4°C.
   - C5 dish: 30 µL goat-anti-mouse IgG + IgM + 10 mL of 50 mM Tris-HCl (pH 9.5)
   - PDGFRβ dish: 30 µL donkey-anti-goat IgG + 10 mL of 50 mM Tris-HCl (pH 9.5)
2. If you plan to plate the isolated pericytes on culture dishes, coat the culture dishes with PDL.
   i. Dilute PDL stock 1:100 in sterile H_2O.
   ii. Add diluted PDL to the dishes: for a 10-cm dish, add 5 mL to each dish; for a six-well plate, add 1 mL per well; and for a 24-well plate, add 250 µL per well.
   iii. Swirl the dishes to evenly coat plastic, and incubate 20–60 min at room temperature.
   iv. Rinse the dishes three times with sterile H_2O, and allow to dry completely before use.
3. If you plan to plate the isolated pericytes on coverslips, coat the coverslips with PDL.
   i. In a Petri dish, rinse ethanol-washed glass coverslips three times with sterile H_2O.
   ii. Suction away the remaining H_2O and separate the coverslips in the dish such that they do not touch each other or the sides of the dish. Allow the coverslips to dry completely (this should take 5–10 min).
   iii. Carefully add 100 µL of diluted PDL to the center of each coverslip; the pDL solution should remain as a bubble on the coverslip. Incubate 30–60 min at room temperature.

Cite this protocol as *Cold Spring Harb Protoc*; doi:10.1101/pdb.prot074955

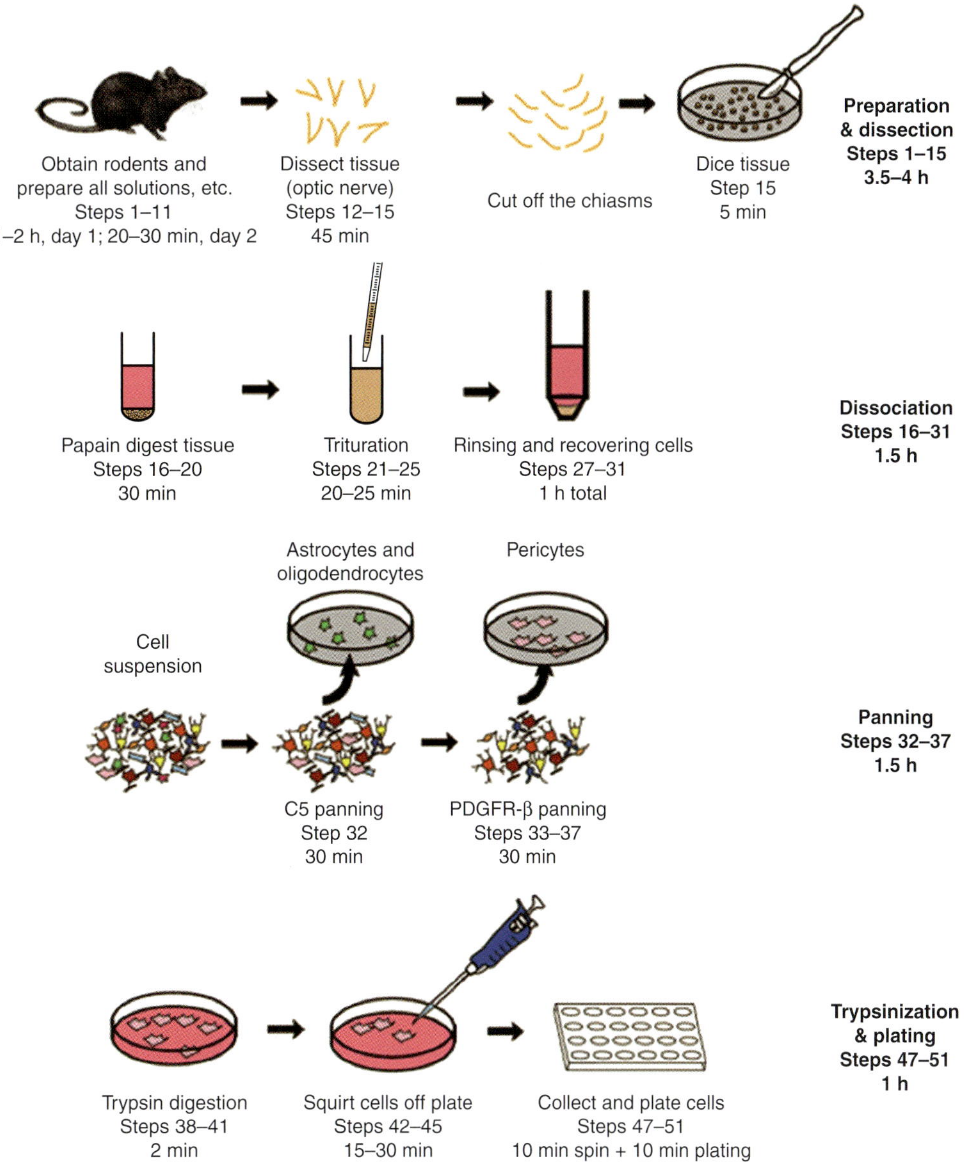

FIGURE 1. Immunopanning of endothelial cells with anti-PDGFRβ.

iv. Rinse the coverslips three times with sterile H_2O.

v. Transfer the coverslips to 24-well tissue culture plates, with one coverslip per well.

vi. Suction any remaining H_2O from the wells, being careful to keep coverslips centered and not touching the sides of the wells. Allow the coverslips to dry completely before use.

4. Prepare pericyte/astrocyte growth medium (PAGM) for optic nerve pericytes.

   i. To 450 mL of DMEM, add 50 mL of FCS, 5 mL of pen-strep, 5 mL of L-glutamine, 5 mL of insulin, and 5 mL of sodium pyruvate.

   ii. Filter the medium into a 500-mL bottle and store at 4°C.

   For this and any other solutions that require filter sterilization, prerinse the filter with 5–10 mL of base liquid (e.g., sterile water or D-PBS) to remove any residual surfactants that may be detrimental to cell health; discard this rinse before filtering the solution to be sterilized.

5. Make 40 mL of 0.2% BSA by combining 38 mL of D-PBS with 2 mL of 4% BSA. Store overnight at 4°C.

Day 2

6. Add PAGM to the culture dishes: for 10-cm dishes, add 10 mL to each plate; for six-well plates, add 1.8 mL per well; and for 24-well plates, add 450 µL per well. Keep the dishes in a 37°C, 10% CO_2 incubator until plating the pericytes in Step 50.
7. Rinse each panning dish from Step 1 three times with PBS, and then coat the dishes with primary antibody solution. Swirl the dishes for even coating, and incubate for ≥2 h at room temperature.
   - C5 dish: 4 mL of C5 hybridoma supernatant
   - PDGFRβ dish: 10 mL of 0.2% BSA + 60 µL each of goat anti-human PDGFRβ and goat-anti mouse PDGFRβ antibody.
8. Prepare the papain solution by combining 85 units of papain with 6 mL of DPBS in a 15-mL conical tube. Incubate the solution in a 37°C water bath to allow the papain to fully dissolve.
9. Equilibrate 10 mL of EBSS (with phenol red) in a 37°C, 10% CO_2 incubator.
10. Add ~300 µL of D-PBS without Ca^{2+}/Mg^{2+} to a 6-cm Petri dish. Add 3 mL of D-PBS without Ca^{2+}/Mg^{2+} to a separate 6-cm Petri dish. These plates will be used during dissection.
11. Prepare the following solutions for dissociation, panning, and trypsinization.
    - Low-ovo inhibitor solution: 9 mL of D-PBS + 1 mL of 10× low-ovomucoid stock solution
    - High-ovo inhibitor solution: 5 mL of D-PBS + 1 mL of 6× high-ovomucoid stock solution
    - Panning buffer: 9 mL D-PBS + 1 mL of 0.2% BSA solution (from Step 6) + 100 µL of 0.5 mg/mL insulin
    - 30% FCS: 14 mL D-PBS + 6 mL heat-inactivated FCS; filter-sterilize through a 0.22-µm filter.

Dissection

12. Kill postnatal day 6–8 Sprague–Dawley rat pups with CO_2 according to appropriate animal protocol.
13. Decapitate each pup with sharp scissors.
14. Open the skull.
    i. Cut the skin along the top midline of the head, and peel back to reveal the skull.
    ii. Insert curved scissors at the opening at the back of the skull, and cut along the left and right edge of the skull, moving from above the ear to just above the eye socket.
    iii. Lift the skull bone to reveal the brain.
15. Isolate the optic nerves.
    i. Cut away the olfactory bulbs at the front of the brain.
    ii. Lift the front of the brain. The optic nerves and optic tracts will be stretching from the base of the brain to the eye sockets. Cut the optic tracts immediately adjacent to the optic chiasm to leave the optic chiasm attached to the two optic nerves.
    iii. Insert the curved scissors into the eye orbit behind the eyeball. Cut at the base of each eyeball to detach the optic nerve from the eyeballs.
    iv. Grab the optic chiasm with forceps. Lift the optic chiasm and the two optic nerves from the base of the skull. Place them in the 6-cm Petri dish with 3 mL of D-PBS without Ca^{2+}/Mg^{2+}.
    v. With the aid of a dissecting microscope, clean any debris away from the nerves, and use a scalpel to separate them from the optic chiasm.
    vi. Move the nerve segments to the 6-cm Petri dish with 300 µL of D-PBS without Ca^{2+}/Mg^{2+}, and dice them with very small dissection scissors.

 Cite this protocol as *Cold Spring Harb Protoc*; doi:10.1101/pdb.prot074955

Dissociation

16. Finish preparing the papain solution.
    i. Add 2 mg of L-cysteine to the papain solution from Step 8. The solution should turn yellow because of increasing acidity.
    ii. Add one drop of NaOH to the solution and swirl. The solution should return to a red color, indicating that the solution has a neutral pH. If it does not, add another drop of NaOH until it does. If you overshoot and the solution turns purple, add drops of HCl until it is neutral.
    iii. Filter the papain solution through 0.22-µm filter to sterilize. Collect in a universal tube.
    iv. Add 100 µL of DNase stock solution.
17. Wash and digest the diced tissue from Step 15 by placing it directly into the universal tube of papain/DNase solution. Incubate the diced optic nerve tissue/papain solution for 45 min in a 35°C water bath. Gently agitate the tissue every 15 min to ensure complete digestion.
18. While the tissue is digesting, add 100 µL of DNase stock solution to the low-ovo inhibitor solution from Step 11. If the low-ovo and high-ovo inhibitor solutions are not already at a neutral pH, add NaOH until they are.
19. When the digestion is complete, allow the tissue pieces to settle. Carefully remove as much papain solution as possible.
20. To stop the papain digestion, gently add 2 mL of low-ovo inhibitor solution from Step 18 to the tissue. Allow the tissue to settle again.
21. Remove and discard the low-ovo inhibitor solution, and add 2 mL of fresh low-ovo inhibitor solution to the tissue.
22. Triturate the tissue by pipetting it six to eight times through a 1-mL pipette tip. Allow the tissue chunks to settle for 1–2 min, and transfer 1–1.5 mL of supernatant to a new 15-mL collection tube labeled "optic nerve." Avoid taking large tissue chunks.

    From this step on, the dissociated cells are in the supernatant, and the chunks of tissue settle to the bottom.
23. Add 1 mL of low-ovo inhibitor solution and repeat Step 22, adding the supernatant to the same "optic nerve" collection tube.
24. Add 1 mL of low-ovo inhibitor solution. Triturate the tissue three times through a 21-gauge needle, allow the tissue chunks to settle for 1–2 min, and transfer the supernatant to the "optic nerve" collection tube.
25. Repeat Step 24, using a 23-gauge needle, until all of the low-ovo inhibitor solution has been used.
26. (*Optional*) Remove a small aliquot (50–100 µL) of the cell suspension from the "optic nerve" collection tube. Add an equal volume of trypan blue to the aliquot, and count the cells with a hemacytometer to monitor viability and effectiveness of dissociation.
27. Pellet the cells in the "optic nerve" collection tube by centrifuging at ~220*g* in a tabletop centrifuge for 15 min at room temperature. Aspirate and discard the supernatant, being careful not to disturb the cell pellet.
28. Resuspend the cells in 6 mL of high-ovo inhibitor solution from Step 11. Immediately centrifuge the cells at ~220*g* in a tabletop centrifuge for 10 min at room temperature.

    Keeping the cells in high-ovo inhibitor solution for an extended period will reduce cell viability.
29. Aspirate and discard the supernatant, being careful not to disturb the cell pellet.
30. Resuspend the cells in 10 mL of panning buffer from Step 11.
31. Incubate the cell suspension for 30 min at 37°C to allow epitope recovery.

    The immunopanning relies on the presence of PDGFRβ on the surface of the pericytes, so it is important to allow enough time after the papain digestion for antigen recovery. After trituration, 30 min of incubation at 37°C is usually sufficient, but recovery time can range from 10 min to 1 h and should be optimized to ensure that enough PDGFRβ is retained at the surface such that pericytes can be fully separated from endothelial cells.

Panning

32. Rinse the C5 panning dish from Step 7 three times with D-PBS, and transfer the cell suspension to the rinsed dish. Incubate the dish for 30 min at room temperature, agitating it at 15-min intervals to ensure that all cells have an opportunity to adhere to its surface.

    In this negative panning step, used to select against cells that adhere to the C5 dish, pericytes should remain in the cell suspension.

33. Rinse the PDGFRβ panning dish three times with D-PBS.
34. Gently shake the C5 dish to loosen nonadherent cells, and transfer the supernatant (the cell suspension) to the PDGFRβ dish. Incubate for 30 min at room temperature, agitating the dish at 15-min intervals to ensure that all cells have an opportunity to adhere to the dish's surface.

    This panning step positively selects for cells that adhere to the PDGFRβ panning dish.

35. Shake the dish to loosen nonadherent cells, and pour off and discard the cell suspension.
36. Rinse the dish six to eight times with D-PBS, shaking the dish during each rinse to loosen nonadherent cells.
37. Before trypsinization, confirm visually under a microscope that nearly all of the nonadherent cells have been removed by rinsing; if not, perform additional rinsing steps.

    There will always be a small number (<0.5%) of floating cells representing dislodged optic nerve pericytes at the end of the rinsing steps.

Trypsinization

38. When the optic nerve pericytes on the PDGFRβ panning dish are in their final D-PBS rinse, prepare trypsin-EBSS by combining 4 mL of equilibrated EBSS (from Step 9) with 100 μL of trypsin stock solution.
39. Remove the D-PBS from the PDGFRβ dish and rinse it with the remaining 6 mL of equilibrated EBSS.
40. Pour off and discard the EBSS, and add 4 mL of trypsin-EBSS solution to the dish. Incubate for 2 min at 37°C.

    Incubation times may vary based on the strength of the lot of trypsin.

41. Stop the trypsin digestion by adding 2 mL of filtered 30% FCS solution (from Step 11) to the dish.
42. Dislodge the optic nerve pericytes from the dish by squirting its surface with a 1-mL pipette tip; squirt once around the entire circumference toward the center of the dish, and then once along the edge of the dish.

    Avoid scraping the pipette tip along the surface of the plate; instead, squirt just above the surface. Avoid generating excess bubbles in the medium during this step.

43. Transfer the cell suspension from the dish to a sterile 50-mL conical tube.
44. Place 5 mL of fresh 30% FCS in the dish. Inspect the dish under a microscope to see if there are particular regions of the plate that still contain adherent cells. Repeat Steps 42 and 43 to dislodge and collect any remaining optic nerve pericytes.

    Adherent cells are often found along the edges or the exact center of the plate.

45. Rinse the plate with the remaining 30% FCS, and add the rinse to the collection tube.
46. Remove a 50–100-μL aliquot of the cell suspension from the collection tube for counting in Step 48.
47. Pellet the optic nerve pericytes at ~220*g* in a tabletop centrifuge for 10 min at room temperature.
48. During the centrifugation, add an equal volume of trypan blue to the aliquot of cells from Step 46. and count the cells with a hemacytometer to monitor the yield.
49. After centrifugation of the cells in Step 47, aspirate and discard the supernatant.

Cite this protocol as *Cold Spring Harb Protoc*; doi:10.1101/pdb.prot074955

Plating

50. Resuspend the optic nerve pericytes in PAGM, and plate them on the tissue culture dishes prepared in Step 6.

    i. To plate the cells on coverslips in multiwell plates, adjust the volume of PAGM in the cell suspension such that 50 µL of medium contains the desired number of cells per well. Place 50 µL of suspended cells onto individual coverslips covered with PAGM; the final volume will be 500 µL/well. Pericytes can be cultured on PDL-coated coverslips without preplating or centrifugation.

    ii. To plate the cells on tissue culture dishes, adjust the volume of PAGM in the cell suspension such that 300 µL of medium contains the desired number of cells per dish. Pipette 300 µL of cell suspension into a 10-cm PDL-coated tissue culture dish containing 10 mL of PAGM. Carefully swirl the dish to evenly distribute the cells.

    Because the cells will divide, confluent monolayers are possible with a wide variety of initial cell densities. Suggested starting points for plating are 1000–10,000 cells per coverslip and 50,000–200,000 cells per 10-cm dish. Pericytes grow well in a variety of media, especially serum-containing media. PDGF-BB can also be used to stimulate growth of pericytes. They have been observed to decrease growth and/or alter their phenotype in the presence of basic fibroblast growth factor.

51. Incubate the optic nerve pericyte cultures in a 37°C, 10% CO_2 incubator. Replace 50% of the medium with fresh medium every 2–3 d.

52. Assess the purity of the cell preparation by staining with a series of markers that positively identify pericytes and negatively identify other cells. Acutely isolated CNS pericytes are PDGFRβ^+, NG2$^+$, RGS5$^+$, claudin-5$^-$, occludin$^-$, CD31$^-$, and VE-cadherin$^-$. Oligodendrocyte progenitor cells (OPCs) are also NG2$^+$, but are PDGFRβ^-, PDGFRα^+. Endothelial cells are CD31$^+$, claudin-5$^+$; and astrocytes are GFAP$^+$, aquaporin 4$^+$. Smooth muscle actin (SMA) immunoreactivity is a good marker of pericytes following a week in culture.

    Contaminating cell types are generally not a problem. Endothelial cells, for instance, do not adhere to PDL-coated coverslips and do not grow as fast as pericytes. Nevertheless, purity should be assessed during each purification. Example images of immunopanned optic nerve pericytes are shown in Figure 2.

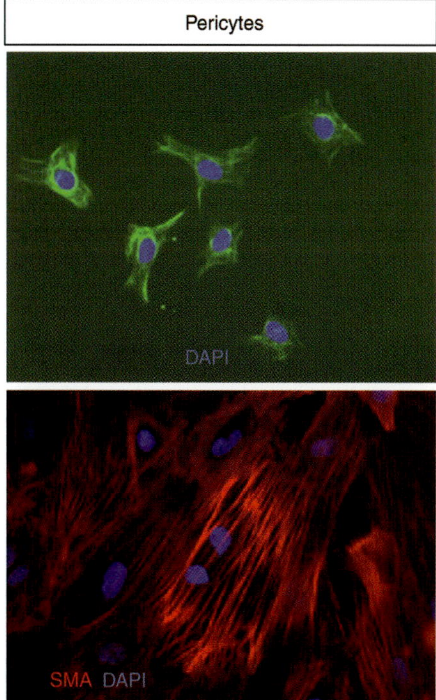

FIGURE 2. Example images of immunopanned optic nerve pericytes. Purified pericytes were stained acutely 2 h after purification with an antibody directed against nestin (green, *top panel*) or following 2 wk in culture with an antibody against smooth muscle actin (red, *bottom panel*), and counterstained with DAPI (blue).

RELATED INFORMATION

For background information on this protocol, see Introduction: Purification and Culture of Central Nervous System Pericytes (Zhou et al.).

RECIPES

Ethanol-Washed Glass Coverslips

Extensively wash 12-mm glass coverslips (Carolina Biological Supply 633029) in 70% ethanol. Perform the washes on a platform shaker in a beaker, with enough motion to lightly agitate the coverslips but not break too many. Wash the coverslips for about 1 mo, exchanging the ethanol approximately every day. (It is fine to skip some exchanges.) Store the washed coverslips in 70% ethanol until use.

High-Ovomucoid Stock Solution (6×)

BSA (Sigma-Aldrich A8806)
D-PBS (Thermo Scientific HyClone SH3026401)
NaOH (1 N)
Trypsin inhibitor (Worthington LS003086)

1. Add 6 g of BSA to 150 mL D-PBS.
2. Add 6 g of trypsin inhibitor and mix to dissolve.
3. Add at least 1.5 mL of 1 N NaOH to adjust the pH; continue adding NaOH as necessary to bring up the pH to 7.4.
4. Bring the volume to 200 mL with D-PBS.
5. Filter-sterilize through a 0.22-µm filter.
6. Make 1.0-mL aliquots and store at −20°C.

Insulin (0.5 mg/mL)

1. Working under a sterile tissue culture hood, prerinse a 0.22-µm filter with sterile H_2O. Discard the flowthrough.
2. Add 50 mg of insulin (either recombinant human insulin [Sigma-Aldrich I2643] or bovine pancreas insulin [Sigma-Aldrich I6634]) to 100 mL of sterile H_2O in a 200-mL beaker. Immediately add 500 µL of 1 N HCl to adjust the pH.
3. Stir the mixture with a sterile pipette to dissolve the insulin completely; avoid creating bubbles.
4. Filter the solution through the prerinsed 0.22-µm filter.
5. Make 5-mL aliquots and store at 4°C for up to 4–6 wk. For use in media, dilute 100× to a concentration of 5 µg/mL.

Low-Ovomucoid Stock Solution (10×)

To prepare, add 3 g of BSA (Sigma-Aldrich A8806) to 150 mL D-PBS. Mix well. Add 3 g of trypsin inhibitor (Worthington LS003086) and mix to dissolve. Add ~1 mL of 1 N NaOH to adjust the pH to 7.4. Bring the volume to 200 mL with D-PBS. Filter-sterilize through a 0.22-µm filter. Make 1.0-mL aliquots and store at −20°C.

Cite this protocol as *Cold Spring Harb Protoc*; doi:10.1101/pdb.prot074955

Poly-D-lysine (PDL) Stock (1 mg/mL)

1. Resuspend poly-D-lysine (PDL; Sigma-Aldrich P6407; molecular weight 70–150 kDa) at 1 mg/mL in borate buffer by combining 50 mg of PDL and 50 mL of 0.15 M boric acid (pH 8.4).
2. Filter to sterilize, then aliquot (e.g., 100 µL/tube). Store aliquots at −20°C.

REFERENCE

Zhou L, Sohet F, Daneman R. 2014. Purification and culture of central nervous system pericytes. *Cold Spring Harb Protoc* doi: 10.1101/pdb.top070888.

CHAPTER 7

Purification and Culture of Central Nervous System Endothelial Cells

Lu Zhou,[1] Fabien Sohet,[2] and Richard Daneman[2,3]

[1]*Department of Neurobiology, Stanford University School of Medicine, Stanford, California 94305-5125;*
[2]*Department of Anatomy, UCSF, San Francisco, California 94143-0452*

Blood vessels are critical for delivering oxygen and nutrients to all tissues in the body. This is especially important in the central nervous system, which is extremely sensitive to hypoxia and ischemia. Blood vessels are made of two main cell types: endothelial cells and mural cells. Endothelial cells form the walls of the blood vessels that generate a lumen through which blood flows. Mural cells are support cells thought to be involved in vessel contractility, vascular remodeling, and regulation of endothelial permeability. On large vessels, including arteries and veins, mural cells are termed vascular smooth muscle cells. On the small vessels of the capillary bed, they are called pericytes. Here, we provide a brief introduction to the methods for purification of endothelial cells, including an immunopanning method that we developed for isolating these cells from the rodent brain and optic nerve.

ENDOTHELIAL CELLS IN THE CENTRAL NERVOUS SYSTEM

Endothelial cells form the walls of the blood vessels, which are critical for delivering blood to all parts of the body, including the central nervous system (CNS). Because of their location, endothelial cells regulate the interaction between the blood and each tissue, including the delivery of nutrients, regulation of clotting, and initiation of immune surveillance. The properties of endothelial cells vary throughout the body, depending on the specific requirements of the tissues they vascularize and the branch of the vascular tree in which they reside (i.e., the arteries, arterioles, capillaries, venules, or veins).

The endothelial cells of the CNS form a physiological structure termed the blood–brain barrier (BBB), which is critical for regulating the interaction between the blood and the nervous tissue. This barrier tightly regulates the neuronal environment by limiting the passive movement of molecules and ions from the blood to the brain and provides the CNS with specific nutrients through selective transport. To accomplish this level of control, CNS endothelial cells differ from endothelial cells in nonneural tissue in that they are held together by tight junctions and possess few transcytotic vesicles; these properties limit the paracellular and transcellular movement of hydrophilic molecules and ions between the blood and the brain (Rubin and Staddon 1999). CNS endothelial cells express a variety of molecular transporters. These transporters deliver specific nutrients down their concentration gradient into the CNS or use ATP to pump out potential toxins up to their concentration gradient (Zlokovic 2008; Hermann and Elali 2012). Although these functions are manifested in the endothelial cells, key transplantation studies have showed that they are not intrinsic to the endothelial cells, but induced by cellular interactions with the neural tissues (Stewart and Wiley 1981). Astrocytes, pericytes, neurons, and neural stem cells have all been implicated in regulating these functions (Janzer and Raff 1987; Rubin et al. 1991b; Weidenfeller et al. 2007).

[3]Correspondence: richard.daneman@ucsf.edu

Cite this introduction as *Cold Spring Harb Protoc;* doi:10.1101/pdb.top070987

Purification and culture of CNS endothelial cells alone or in coculture with other neural cells allow for examination of the cellular and molecular mechanisms that regulate barrier properties of the endothelial cells. For instance, astrocyte–endothelial cell cocultures have been used to show that close apposition of astrocytes to endothelial cells enhances the electrical resistance of the endothelial cell tight junctions (Beck et al. 1984; Rubin et al. 1991a).

Formation of blood vessels (vasculogenesis and angiogenesis) is critical for both the development of tissues and the progression of tumors (Carmeliet 2003), and purification of endothelial cells from many tissues has facilitated studies of the molecular mechanisms of these processes. Endothelial cells have also been implicated in modulating the development and function of tissues throughout the body, and purification and culture of CNS endothelial cells has been used to identify their role in development of the brain. For instance, purified endothelial cells have been shown to regulate astrocyte differentiation, neuronal survival, and neural stem cell self-renewal and neurogenesis (Mi et al. 2001; Shen et al. 2004; Dugas et al. 2008).

PURIFICATION OF ENDOTHELIAL CELLS

Several different techniques have been used to purify endothelial cells from the CNS. One method involves homogenization of tissue, followed by isolation of microvessels by centrifugation through dextran or sucrose gradients, and then isolation of endothelial cells by selective adherence to specific substrates. This general procedure has been used to purify endothelial cells from the brains of mice, rats, cows, pigs, and humans. An alternative approach has been to purify the endothelial cells by using magnetic beads coupled to anti-CD31 antibodies (van Beijnum et al. 2008; Springhorn 2011). In many cases, purification has been coupled with culturing the cells in puromycin for a limited period (Perrière et al. 2005). Puromycin is toxic to the cells, but is pumped out by efflux transporters specifically expressed by CNS endothelial cells, so addition of puromycin to cell culture medium kills contaminating nonendothelial cells. Puromycin treatment has proven effective with cells from adult specimens, but the efflux transporters have lower activity at earlier developmental time periods, so it is not efficient to select endothelial cells from embryonic or neonatal rodent brains. We have achieved extremely pure endothelial cell populations by using fluorescence-activated cell sorting (FACS) with a Tie2GFP transgenic mouse line (Daneman et al. 2010). FACS, however, is expensive and time consuming. To avoid the need for FACS, we have developed a method for purifying CNS endothelial cells by immunopanning, as described in Protocol 1: Purification of Endothelial Cells from Rodent Brain by Immunopanning (Zhou et al.). This prospective isolation method, based on the binding of cells to a dish coated with anti-CD31 antibodies, produces pure populations of CNS endothelial cells that can be analyzed acutely or cultured.

REFERENCES

Beck DW, Vinters HV, Hart MN, Cancilla PA. 1984. Glial cells influence polarity of the blood–brain barrier. *J Neuropathol Exp Neurol* **43:** 219–224.

Carmeliet P. 2003. Angiogenesis in health and disease. *Nat Med* **9:** 653–660.

Daneman R, Zhou L, Agalliu D, Cahoy JD, Kaushal A, Barres BA. 2010. The mouse blood–brain barrier transcriptome: A new resource for understanding the development and function of brain endothelial cells. *PLoS ONE* **5:** e13741.

Dugas JC, Mandemakers W, Rogers M, Ibrahim A, Daneman R, Barres BA. 2008. A novel purification method for CNS projection neurons leads to the identification of brain vascular cells as a source of trophic support for corticospinal motor neurons. *J Neurosci* **28:** 8294–8305.

Hermann DM, Elali A. 2012. The abluminal endothelial membrane in neurovascular remodeling in health and disease. *Sci Signal* **5:** re4.

Janzer RC, Raff MC. 1987. Astrocytes induce blood–brain barrier properties in endothelial cells. *Nature* **325:** 253–257.

Mi H, Haeberle H, Barres BA. 2001. Induction of astrocyte differentiation by endothelial cells. *J Neurosci* **21:** 1538–1547.

Perrière N, Demeuse P, Garcia E, Regina A, Debray M, Andreux JP, Couvreur P, Scherrmann JM, Temsamani J, Couraud PO, et al. 2005. Puromycin-based purification of rat brain capillary endothelial cell cultures. Effect on the expression of blood–brain barrier–specific properties. *J Neurochem* **93:** 279–289.

Rubin LL, Staddon JM. 1999. The cell biology of the blood–brain barrier. *Annu Rev Neurosci* **22:** 11–28.

Rubin LL, Barbu K, Bard F, Cannon C, Hall DE, Horner H, Janatpour M, Liaw C, Manning K, Morales J, et al. 1991a. Differentiation of brain endothelial cells in cell culture. *Ann NY Acad Sci* **633:** 420–425.

Cite this introduction as *Cold Spring Harb Protoc*; doi:10.1101/pdb.top070987

Rubin LL, Hall DE, Porter S, Barbu K, Cannon C, Horner HC, Janatpour M, Liaw CW, Manning K, Morales J. 1991b. A cell culture model of the blood–brain barrier. *J Cell Biol* **115:** 1725–1735.

Shen Q, Goderie SK, Jin L, Karanth N, Sun Y, Abramova N, Vincent P, Pumiglia K, Temple S. 2004. Endothelial cells stimulate self-renewal and expand neurogenesis of neural stem cells. *Science* **304:** 1338–1340.

Springhorn JP. 2011. Isolation of human capillary endothelial cells using paramagnetic beads conjugated to anti-PECAM antibodies. *Cold Spring Harb Protoc* doi: 10.1101/pdb.prot4479.

Stewart PA, Wiley MJ. 1981. Developing nervous tissue induces formation of blood–brain barrier characteristics in invading endothelial cells: A study using quail-chick transplantation chimeras. *Dev Biol* **84:** 183–192.

van Beijnum JR, Rousch M, Castermans K, van der Linden E, Griffioen AW. 2008. Isolation of endothelial cells from fresh tissues. *Nat Protoc* **3:** 1085–1091.

Weidenfeller C, Svendsen CN, Shusta EV. 2007. Differentiating embryonic neural progenitor cells induce blood–brain barrier properties. *J Neurochem* **101:** 555–565.

Zhou L, Sohet F, Daneman R. 2014. Purification of endothelial cells from rodent brain by immunopanning. *Cold Spring Harb Protoc* doi: 10.1101/pdb.prot074963.

Zlokovic BV. 2008. The blood-brain barrier in health and chronic neurodegenerative disorders. *Neuron* **57:** 178–201.

Protocol 1

Purification of Endothelial Cells from Rodent Brain by Immunopanning

Lu Zhou,[1] Fabien Sohet,[2] and Richard Daneman[2,3]

[1]*Department of Neurobiology, Stanford University School of Medicine, Stanford, California 94305-5125;*
[2]*Department of Anatomy, UCSF, San Francisco, California 94143-0452*

This protocol describes the use of immunopanning to purify endothelial cells from the rodent brain. Immunopanning permits the prospective isolation of endothelial cells from nervous tissue by relying on the binding of the endothelial cells to an anti-CD31 antibody adhered to a Petri dish. The cells are viable at the end of this gentle procedure, and they can be analyzed acutely for gene expression or cultured alone or in coculture with other central nervous system (CNS) cell types, including CNS pericytes and CNS astrocytes. This procedure can be used to isolate endothelial cells from either rat or mouse. We have suggested specific antibodies that work for each species. Note that endothelial cells from rats and mice have different morphologies; in general, rat CNS endothelial cells are longer and thinner than mouse CNS endothelial cells. This procedure can also be used to purify endothelial cells from different regions of the CNS, including brain and optic nerve. Dissociation procedures must be optimized for each tissue.

MATERIALS

It is essential that you consult the appropriate Material Safety Data Sheets and your institution's Environmental Health and Safety Office for proper handling of equipment and hazardous materials used in this protocol.

RECIPES: Please see the end of this protocol for recipes indicated by <R>. Additional recipes can be found online at http://cshprotocols.cshlp.org/site/recipes.

Reagents

Antibodies and reagents for immunostaining and identifying endothelial cells and contaminating cells (e.g., antibodies against claudin 5 and occludin, DAPI; see Step 56)

Antibodies for immunopanning (mouse prep)

- Secondary antibody: goat anti-rat IgG (H + L) (Jackson ImmunoResearch 112-005-167), 1.2 mg/mL
- Primary antibody, negative selection: rat anti-mouse CD45 (AbD Serotec MCA1301GA), 0.1 mg/vial
- Primary antibody, positive selection: rat anti-mouse CD31 (BD Pharmingen 553370), 0.5 mg/mL

Antibodies for immunopanning (rat prep)

- Secondary antibody: goat anti-mouse IgG + IgM (H + L) (Jackson ImmunoResearch 115-005-044), 2.3 mg/mL

[3]Correspondence: richard.daneman@ucsf.edu

Cite this protocol as *Cold Spring Harb Protoc*; doi:10.1101/pdb.prot074963

Primary antibody, negative selection: mouse anti-rat CD45 (AbD Serotec MCA589), 2 mL/vial
Primary antibody, positive selection: mouse anti-rat CD31 (Fitzgerald 10R-CD31gRT), 1 mg/mL

Bovine serum albumin (BSA), 4%

To prepare 4% BSA in Dulbecco's phosphate-buffered saline (D-PBS), dissolve 8 g of BSA (Sigma-Aldrich A4161) in 150 mL of D-PBS (Invitrogen 14287-080). Adjust the pH to 7.4 with 1 N NaOH, and bring the volume to 200 mL. Filter through a prerinsed 0.22-µm filter, and store in 1.0-mL aliquots at −20°C.

Collagen IV

Dilute mouse collagen IV (BD Biosciences 354233) to 100 µg/mL in neurobasal medium. Vortex the tube and then filter the solution with a 0.4-µm syringe filter to remove the precipitate. Aliquot and store at −80°C.

DNase stock solution (12,5000 U/mL)

To prepare DNase stock solution in Earle's balanced salt solution (EBSS), dissolve 12,500 U of DNase I (Worthington LS002007) in 1 mL of ice-cold EBSS. Filter-sterilize the solution through a prerinsed 0.22-µm filter on ice, and freeze 200-µL aliquots overnight at −20°C. Store at −20°C to −30°C.

D-PBS (Invitrogen 14287-080), with and without phenol red

To help ensure that the D-PBS is at a neutral pH, add the indicator phenol red (Sigma-Aldrich P0290) at a 1:1000 dilution from 0.5% stock.

D-PBS without Ca^{2+}/Mg^{2+} (Invitrogen 14190-144)

D-PBS without Ca^{2+}/Mg^{2+} is used ONLY for tissue dissection. Cells will not stick to the panning dishes in Ca^{2+}/Mg^{2+}-free D-PBS. Do not add phenol red.

Earle's Buffered Saline Solution (EBSS, Invitrogen 14155-063)

To help ensure that the EBSS is at a neutral pH, add the indicator phenol red (Sigma-Aldrich P0290) at a 1:1000 dilution from 0.5% stock.

Endothelial cell growth medium <R>

Prepare this medium on the second day of the purification procedure (see Step 3).

Ethanol (70%)
Ethanol-washed glass coverslips <R>
Fetal calf serum (FCS), heat-inactivated

Make 50-mL aliquots of FCS (Invitrogen 16000-036) and heat-inactivate for 30 min in a water bath at 55°C. Store the aliquots at −20°C.

High-ovomucoid (high-ovo) stock solution (6×) <R>

Prepare with D-PBS with Ca^{2+}/Mg^{2+} (Invitrogen 14287-080).

Inhibitor stock solution <R>
Insulin (0.5 mg/mL) <R>
L-Cysteine (Sigma-Aldrich C-7477)
Low-ovomucoid (low-ovo) stock solution (10×) <R>

Prepare with D-PBS with Ca^{2+}/Mg^{2+} (Invitrogen 14287-080).

Mouse or rat

The purification procedure as it appears here was developed for P20 rodents, but it can be used to isolate endothelial cells from the brain of any age mouse or rat. Some modifications (e.g., the duration of trypsinization) will be necessary when purifying CNS endothelial cells from rodents of different ages.

Neurobasal medium (Gibco 21103-049)
Papain (Worthington Biomedical LS003126)
Papain buffer <R>
Penicillin/streptomycin (Pen-Strep, Gibco 15140-122)

Prepare 5-mL aliquots of this 100× solution and store at 4°C.

Phosphate-buffered saline (PBS)
Poly-D-lysine (PDL) stock (1 mg/mL) <R>
Tris-HCl (50 mM, pH 9.5)
Trypan blue (Invitrogen 15250-061)
Trypsin stock solution

Dissolve the trypsin (Sigma-Aldrich T9935) at 50,000 U/mL in EBSS on ice. If the trypsin does not dissolve easily, try stirring it with a magnetic stir bar or altering the pH with NaOH or HCl. When fully dissolved, store 200-µL aliquots at −80°C.

Equipment

CO_2 chamber and tank for killing rats
Conical tubes (15 and 50 mL)
Dissection equipment
- Biohazard disposal bag
- Dissection microscope
- Forceps (#5 or #55)
- Large, sharp scissors for decapitation (e.g., Roboz RS-6820)
- Scalpel with #10 blade
- Small, curved scissors (e.g., Roboz RS-5675)

Equilibration setup (a source of carbon dioxide [5% CO_2/95% O_2] with a line leading to a sterile hood)
Filter sterilization system with 250-mL bottle
Heat block base with a flat insert, preset to 34°C in a sterile hood
Hemacytometer
Incubator at 37°C, 10% CO_2
Microscope, compound
Nitex mesh filter (20 µm) (Small Parts B0015H4H1A)

Cut the mesh into 4-in. squares and autoclave.

Paper towels
Petri dishes (6, 10, and 15 cm)
Petri dish lid (6 cm) with a hole in the center that accommodates a 0.22-µm filter

Use flamed forceps (spray with ethanol and use a Bunsen flame to sterilize) to melt the hole into the center of the lid.

Pipette (1 mL)
Pipettes, serological (2, 5, and 10 mL)
Pipettor, powered (e.g., Pipet-Aid)
Refrigerator preset to 4°C
Sterile laminar flow hood for tissue culture, equipped with aspirator
Syringe filters (0.22 µm)
Table top centrifuge (with 15- and 50-mL tube adaptors) at room temperature
Tissue culture dishes (24 well)
Water baths, preset to 35°C and 37°C

METHOD

This method begins with isolation of mouse or rat brain and creation of a single-cell suspension of CNS cells. These cells then undergo two immunopanning steps to purify the endothelial cells before they are plated (see Fig. 1 for an overview). The entire protocol is a 2-d procedure; Steps 1 and 2 are performed on the first day. Perform all steps in a sterile tissue culture hood, with the exception of the dissection. All solutions in contact with cells must be sterile. One set of panning plates and reagents is sufficient for one postnatal rat or mouse brain.

Preparation of Panning Dishes, Coverslips, and Solutions

Day 1

1. Prepare panning dishes by coating Petri dishes with secondary antibody solution. Swirl the dishes until their surfaces are evenly coated, and incubate them overnight at 4°C.

Cite this protocol as *Cold Spring Harb Protoc*; doi:10.1101/pdb.prot074963

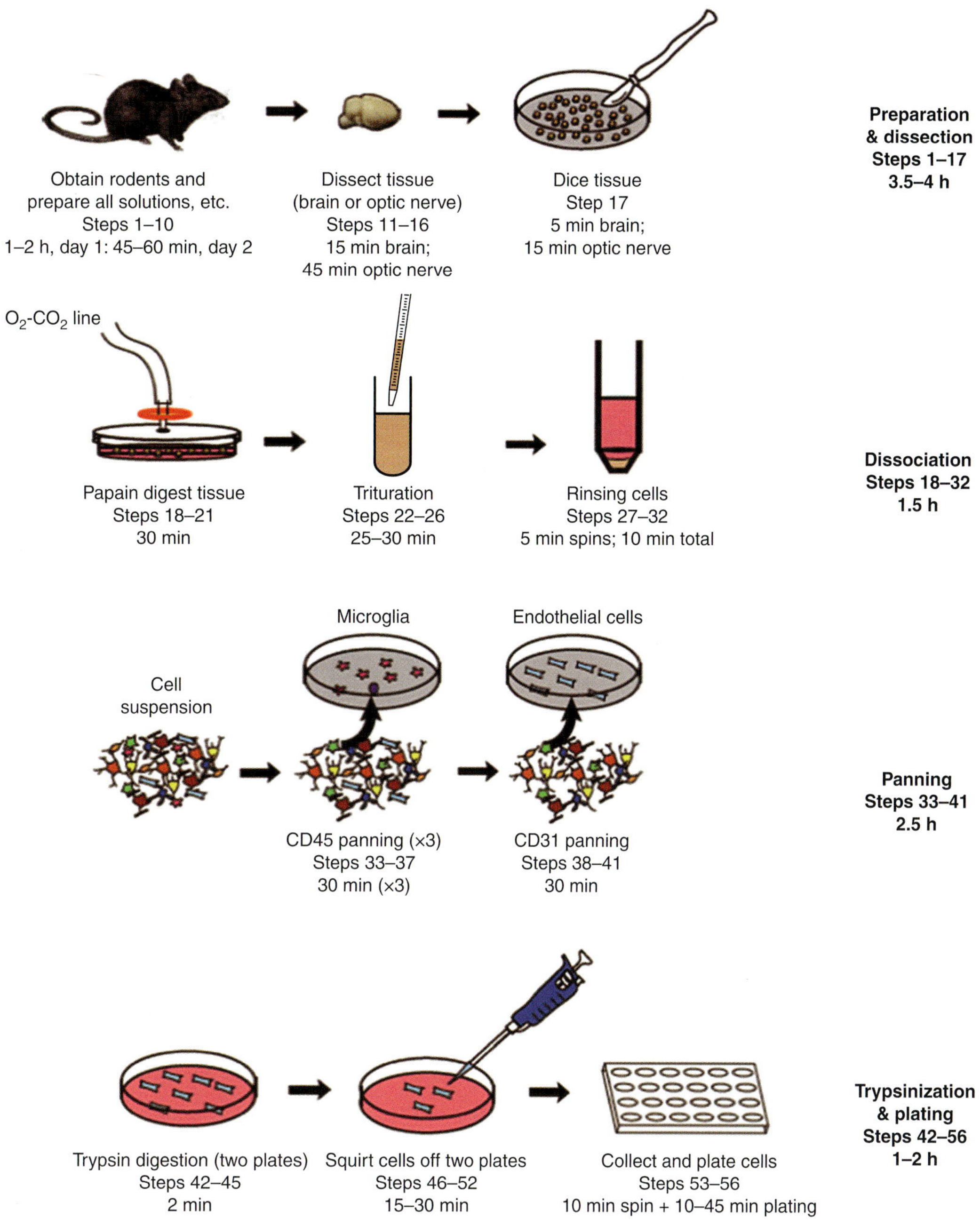

FIGURE 1. Immunopanning of endothelial cells with anti-CD31.

i. Prepare three 15-cm Petri dishes for negative selection:
   - Rat prep: 50 µL goat anti-mouse IgG + IgM (H + L) + 15 mL of 50 mM Tris-HCl (pH 9.5) per dish
   - Mouse prep: 50 µL goat anti-rat IgG (H + L) + 15 mL of 50 mM Tris-HCl (pH 9.5) per dish

ii. Prepare two 10-cm Petri dish for positive selection:
   - Rat prep: 30 µL goat anti-mouse IgG + IgM (H + L) + 10 mL of 50 mM Tris-HCl (pH 9.5) per dish
   - Mouse prep: 30 µL goat anti-rat IgG (H + L) + 10 mL of 50 mM Tris-HCl (pH 9.5) per dish

2. Prepare coverslips.
   i. Dilute PDL stock 1:100 in sterile H_2O.
   ii. In a 10-cm Petri dish, rinse 15–20 ethanol-washed glass coverslips three times with sterile H_2O.
   iii. Aspirate the remaining H_2O and separate the coverslips in the dish such that they do not touch each other or the sides of the dish. Allow the coverslips to dry completely (this should take 5–10 min).
   iv. Carefully add 100 µL of diluted PDL to the center of each coverslip; the PDL solution should remain as a bubble on the coverslip. Incubate 20–60 min at room temperature.
   v. Rinse the coverslips three times with sterile H_2O and aspirate until dry.
   vi. Prepare collagen IV.
   vii. Add 100 µL of collagen IV to each slip and incubate overnight at 37°C.

      The collagen solution can be left on the slips for 1–3 d. Longer incubation time leads to better adhesion of the cells.

Day 2

3. Make solutions.
   i. Prepare two tubes of 0.2% BSA, each with 2 mL of 4% BSA + 38 mL of D-PBS.
   ii. Make panning buffer by combining 18 mL of D-PBS with 2 mL of 0.2% BSA and 200 µL of 0.5 mg/mL insulin.
   iii. Prepare 30% FCS by adding 50 mL of FCS to 115 mL of D-PBS. Filter the solution in a 250-mL bottle and label "30% FCS."
   iv. Place 20 mL of EBSS with 1× phenol red in a 37°C, 10% CO_2 incubator to equilibrate it for trypsinization.
   v. Prepare endothelial cell growth medium (see Reagents).

      For any solutions that require filter sterilization, prerinse the filter with 5–10 mL of base liquid (e.g., D-PBS or sterile H_2O) to remove any residual surfactants that may be detrimental to cell health; discard this rinse before filtering the solution to be sterilized.

4. Rinse each panning dish from Step 1 three times with PBS, and then coat the dishes with primary antibody solution. Swirl the dishes for even coating, and incubate for ≥2 h at room temperature.
   i. Prepare the three 15-cm Petri dishes for negative selection:
      - Rat prep: 45 µL mouse anti-rat CD45 + 15 mL 0.2% BSA (from Step 3) per dish
      - Mouse prep: 10 µL rat anti-mouse CD45 + 15 mL 0.2% BSA per dish
   ii. Prepare the two 10-cm Petri dishes for positive selection:
      - Rat prep: 40 µL mouse anti-rat CD31 + 10 mL 0.2% BSA per dish
      - Mouse prep: 40 µL rat anti-mouse CD31 + 10 mL 0.2% BSA per dish
5. Prepare equipment for enzymatic dissociation.
   i. Presterilize a 6-cm dish lid with a central hole by spraying with 70% ethanol. Allow the lid to dry in a sterile hood.
   ii. Place a wet paper towel on a heating block base set to 34°C in the hood.
   iii. Attach a 0.22-µm filter to the end of a 5% CO_2/95% O_2 supply line, and put the sterile end of the filter through the hole in the sterilized 6-cm Petri dish lid.
6. Wash the coverslips (from Step 2) three times with neurobasal medium to remove excess collagen IV. Use sterile forceps to transfer the coverslips to a 24-well plate, and cover the slips with neurobasal or endothelial cell growth medium. Leave the plate in the hood.

Cite this protocol as *Cold Spring Harb Protoc*; doi:10.1101/pdb.prot074963

7. Prepare the dissecting area.
   i. Spray the dissection tools with 70% ethanol and let dry.
   ii. Place a biohazard disposal bag in the dissection area.
8. Prepare and equilibrate solutions.
   i. Put 10 mL of enzyme stock solution (papain buffer) in a 50-mL conical tube labeled "papain."
   ii. Put 21 mL of inhibitor stock solution into each of two 50-mL conical tubes labeled "low-ovo."
   iii. Put 10 mL of inhibitor stock solution into a 50-mL conical tube labeled "high-ovo."
   iv. Break a 2-mL pipette, attach a 0.22-µm filter, and attach it to the 5% CO_2/95% O_2 line. Bubble 5% CO_2/95% O_2 through each solution for 5–10 min until its color changes to a pinkish red, and then cap the tubes.
   v. Put the "papain" tube in a 35°C water bath, and leave the "low-ovo" and "high-ovo" tubes at room temperature.
9. Prepare dishes for dissection: Add 10 mL of D-PBS without Ca^{2+}/Mg^{2+} to a 6-cm Petri dish and 0.5 mL of D-PBS without Ca^{2+}/Mg^{2+} to a second 6-cm dish.
10. Prepare the papain solution.
    i. Add 50 units of papain to the equilibrated papain buffer in the "papain" tube.
    ii. Add 2 mg of L-cysteine to the solution, and swirl the tube.
    iii. Place the tube in a 37°C water bath to allow the papain to dissolve.

Dissection

11. Kill the rat or mouse with CO_2 according to appropriate animal protocol. Decapitate with large scissors.
12. Insert small, curved scissors at the opening at the base of the skull where the spinal cord exits. Cut around the midline of the skull, moving from just above the ears to just above the eyes.
13. Lift the top of the skull from its posterior end to reveal the brain, which should remain in the base of the skull.
14. Use the curved scissors to remove the olfactory bulbs at the front of the forebrain.
15. (*Optional*) Insert the scissors over the cerebellum toward the midbrain and cut away the hindbrain, cerebellum, and midbrain structures.
16. Use the scissors and forceps to lift the brain out of base of the skull. Transfer the brain to the 6-cm Petri dish containing 10 mL of D-PBS without Ca^{2+}/Mg^{2+}. With the aid of a dissecting microscope, remove any remaining unwanted tissues, including the meninges.
17. Place the remaining brain tissue into the 6-cm Petri dish containing 0.5 mL of D-PBS without Ca^{2+}/Mg^{2+}. Dice the tissue into ~1 mm^2 pieces with a #10 scalpel blade.

Dissociation

18. Sterilize the papain solution by filtering it through a 0.22-µm filter, and then add the solution directly to the diced tissue in the 6-cm Petri dish. Add 200 µL of DNase stock solution to the dish.
19. Place the Petri dish with the tissue and papain solution in the empty 34°C heat block. Cover the dish with the cleaned lid from Step 5 and keep the tissue under 5% CO_2/95% O_2 gas flow for 30 min, gently agitating the tissue every 15 min to ensure complete tissue digestion.

Digestion of the brain at 34°C slightly increases the health of the cells obtained (relative to digestion at 37°C).

The papain-mediated digestion step is critical for this purification. Recovery of antigens to the surface of endothelial cells after trituration has not proven exceptionally useful, so it is important to optimize the papain digestion procedure itself. Overdigestion leads to cleavage of the CD31 antigen from the surface, whereas underdigestion leads to incomplete dissociation; both result in reduced cell recovery. The amount of papain and the incubation time should be optimized for each batch of papain and when purifying cells from different amounts of tissues or from rodents of different ages.

20. While the tissue is digesting, add 1.5 mL of low-ovomucoid stock solution and 100 µL of DNAse to each "low-ovo" tube from Step 8. Add 2 mL of high-ovomucoid stock solution and 50 µL of DNAse to the "high-ovo" tube. If the ovomucoid solutions are prepared early, blow 5% CO_2/ 95% O_2 through a 22-µm filter over (not in) the solutions to equilibrate.

    Dissociation is performed in low-ovo solution because the cells survive trituration better in a lower protein solution; high-ovo solution is subsequently used to fully quench any residual papain enzymatic activity.

21. When the digestion is complete, transfer the papain solution with the tissue to a sterile, 50-mL conical tube and allow the tissue pieces to settle. Carefully remove as much papain solution as possible.
22. To stop the papain digestion, gently add 4 mL of low-ovo inhibitor solution from Step 20 to the tissue. Allow the tissue to settle again.
23. Remove and discard the low-ovo inhibitor solution. Repeat this washing step twice.
24. Add 10 mL of low-ovo solution to a 50-mL conical tube labeled "P20 cortex."
25. Add 4 mL of fresh low-ovo inhibitor solution to the tissue. Triturate the tissue by gently pipetting it six to eight times through a 5-mL pipette. Allow the tissue chunks to settle for 1–2 min, and transfer 1–1.5 mL of supernatant to the "P20 cortex" collection tube. Avoid taking large tissue chunks.

    From this step on, the dissociated cells are in the supernatant, and the chunks of tissue settle to the bottom.

26. Repeat the trituration (Step 25) four to five times, increasing the vigor of trituration each time until the tissue is completely dissociated. Continue adding the cell supernatant to the "P20 cortex" collection tube.
27. Allow the dissociated tissue in the "P20 cortex" tube to settle for 3–5 min. Remove any clumps from the bottom with a 2-mL pipette.
28. Carefully layer 12 mL of high-ovo inhibitor solution under the dissociated cells.
29. Centrifuge the cells at 136*g* in a tabletop centrifuge for 5 min at room temperature.

    The cells will move through the high-ovo inhibitor solution as they form a pellet.

30. Aspirate and discard the supernatant, being careful not to disturb cell pellet. Resuspend the cells in 15 mL of panning buffer from Step 3.
31. (*Optional*) Remove a small aliquot (50–100 µL) of cell suspension from the "P20 cortex" tube. Add an equal volume of trypan blue to the aliquot, and count the cells with a hemacytometer to monitor viability and effectiveness of dissociation.
32. Form a cone from a sterile Nitex filter. Place it over a 50-mL conical tube and prewet it with 2 mL of panning buffer. Slowly filter the dissociated cells to separate the single cells from any remaining clumps.

Panning

33. Rinse a negative selection panning dish from Step 4 three times with D-PBS, and transfer the cell suspension to the rinsed dish. Incubate for 30 min at room temperature, agitating the dish at 15-min intervals to ensure that all cells have an opportunity to adhere to the dish's surface.

 Cite this protocol as *Cold Spring Harb Protoc*; doi:10.1101/pdb.prot074963

In these negative panning steps, unwanted cells will adhere to the dish, and the endothelial cells should remain in the cell suspension.

34. Rinse a second negative selection panning dish from Step 4 three times with D-PBS.
35. Gently shake the first dish to loosen nonadherent cells, and transfer the supernatant (the cell suspension) to the second dish. Incubate for 30 min at room temperature, agitating the dish at 15-min intervals.
36. Rinse the third negative selection panning dish from Step 4 three times with D-PBS.
37. Gently shake the second dish to loosen nonadherent cells, and transfer the cell suspension to the third dish. Incubate for 30 min at room temperature, agitating the dish at 15-min intervals.
38. Rinse the two positive selection panning dishes from Step 4 three times with D-PBS.
39. Shake the third negative selection dish to loosen nonadherent cells, and transfer the cell suspension to the two positive selection panning dishes (half of the suspension to each dish). Incubate for 30 min at room temperature, agitating the dishes at 15-min intervals.

    This positive panning step selects for endothelial cells, which adhere to the anti-CD31-coated dishes.
40. Pour off and discard the supernatant. Rinse the dishes six to eight times with D-PBS to wash off nonadherent cells.
41. Before trypsinization, confirm visually under a microscope that nearly all nonadherent cells have been removed by rinsing; if not, perform additional rinsing steps.

    There will always be a small number (<0.5%) of floating cells representing dislodged endothelial cells at the end of the rinsing steps.

Trypsinization

42. When the endothelial cells are in their final D-PBS rinse, prepare trypsin–EBSS by combining 8 mL of equilibrated EBSS (from Step 3) with 100 µL of trypsin stock solution.
43. Remove the D-PBS from the CD31 dishes. Rinse each dish with 6 mL of equilibrated EBSS.
44. Pour off and discard the EBSS, and add 4 mL of trypsin–EBSS solution to each dish. Incubate for 2 min in a 37°C incubator.
45. Stop the trypsin digestion by adding 1 mL of 30% FCS (from Step 3) to each dish.
46. Dislodge the endothelial cells from each dish by squirting its surface with a 1-mL pipette tip; squirt once around the entire circumference toward the center of the dish, and then once along the edge of the dish.

    Avoid scraping the pipette tip along the surface of the plate; instead, squirt just above the surface. Avoid generating excess bubbles in the medium during this step.
47. Transfer the cell suspension from the dishes to a sterile 15-mL conical tube.
48. Add 5 mL of fresh 30% FCS to each dish. Inspect the dishes under a microscope to see whether there are particular regions of the plates that still contain adherent cells. Repeat Steps 46 and 47 to dislodge and collect any remaining cells.

    Adherent cells are often found along the edges or the exact center of the plate.
49. Rinse the dishes with the remaining 30% FCS, and add the rinse to the collection tube.
50. Remove a 50- to 100-µL aliquot of the cell suspension from the collection tube for counting in Step 52.
51. Pellet the cells at ~220*g* in a tabletop centrifuge for 10 min at room temperature.
52. During the centrifugation, add an equal volume of trypan blue to the aliquot of cells from Step 50, and count the cells with a hemacytometer to monitor the yield.

One P20 rat brain should yield ~2 × 10^6 endothelial cells, and one P60 mouse brain should yield ~1–1.5 × 10^6 endothelial cells.

Plating

53. After centrifugation of the cells in Step 51, aspirate and discard the supernatant. Resuspend the cell pellet in endothelial cell growth medium such that 50 µL of medium contains the desired number of cells per well (typically 50,000 endothelial cells per coverslip in a 24-well plate).
54. Preplate the cells.
    i. Completely remove the medium from the collagen IV-coated coverslips prepared in Step 6 and place a 50-µL spot of suspended cells at the center of each coverslip.
    ii. Incubate the cells for 20–45 min in a 37°C, 10% CO_2 incubator to allow the cells to adhere to the coverslips.
55. Add 500 µL of the desired medium to each well. Incubate the endothelial cell cultures in a 37°C, 10% CO_2 incubator. Replace 50% of the medium with fresh medium every 2 d.

    Endothelial cells can be cultured in a variety of media. bFGF is an excellent mitogen. Serum can be used to promote health as well, but the concentration should be optimized to ensure that contaminating cells do not proliferate. We favor 0.5% serum for healthy, pure cultures. Puromycin can be added to the media for the first 1–3 d of culture to kill contaminating cells that do not express P-glycoprotein efflux transporters.

    See Troubleshooting.
56. Assess the purity of the cell preparation by staining with a series of markers that positively identify endothelial cells and negatively identify other cells. Acutely isolated CNS endothelial cells are claudin 5^+, $occludin^+$, $CD31^+$, VWF^+, and $VE\text{-}Cadherin^+$. Acutely purified pericyte contaminants are $PDGFR\beta^+$, $NG2^+$, claudin 5^-, and $occludin^-$. After culture, endothelial cells can often alter their antigens, including tight junction proteins. CD31 immunoreactivity remains an excellent identifier of endothelial cells in culture. SMA and PDGFRβ immunoreactivity are good markers of pericytes in culture. Examples of immunopanned endothelial cells are shown in Figure 2.

    See Troubleshooting.

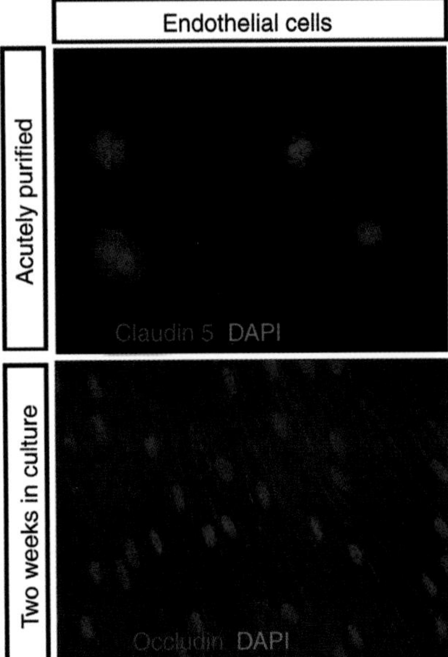

FIGURE 2. Immunopanned endothelial cells. Purified endothelial cells were stained with an antibody directed against claudin 5 acutely (2 h) after purification (red, *top* panel) or with an antibody against occludin after 2 wk in culture (red, *bottom* panel) and counterstained with DAPI (blue, both panels).

Cite this protocol as *Cold Spring Harb Protoc*; doi:10.1101/pdb.prot074963

TROUBLESHOOTING

Problem (Step 55): Endothelial cells do not adhere to the coverslips after plating.

Solution: Adherence of endothelial cells to the substrate can be quite tricky. In our experience, endothelial cells adhere to collagen IV–coated coverslips but not coverslips coated with vitronectin, fibronectin, laminin, or other collagens. Mixtures of matrices can be used as long as they contain collagen IV. In addition, cells should be preplated for at least 20 min, otherwise they tend not to adhere to the coverslips. Alternatively, the cells can be centrifuged in the dish to increase adherence to the coverslips. This method is not as effective as preplating.

Problem (Step 56): Endothelial cell cultures are contaminated by other cells.

Solution: Contaminating cells can be a problem if they outgrow endothelial cells. Microglial cells can be problematic, especially in preparations of tissues taken from P5 to P15 rodents. Contamination can be avoided by increasing the number of CD45 depletion dishes or isolating cells from younger or older rodents. Mature CNS endothelial cells express high levels of p-glycoprotein, a multidrug-resistant pump that makes them insensitive to certain drugs, including puromycin. Therefore, puromycin can be added to cell medium to kill contaminating cell types. The amount of p-glycoprotein expressed by endothelial cells increases during development, and therefore this step is used only for mature endothelial cells (>P20), and the concentration and duration of puromycin addition should be optimized. Contamination also occurs when pericytes adhere to endothelial cells; this problem can be alleviated by including a series of PDGFRβ depletion dishes or optimizing the puromycin concentration.

RELATED INFORMATION

For background information on this protocol, see Introduction: Purification and Culture of Central Nervous System Endothelial Cells (Zhou et al.).

RECIPES

bFGF (50 µg/mL)

Add 1 mL of 5 mM Tris (pH 7.6)/0.1% BSA to a 50-µg vial of bFGF powder (recombinant human FGF-basic; PeproTech 100-18B). Aliquot the solution. Store the aliquots at −20°C to −30°C. Dilute 1000× for use in cell medium.

EBSS Stock (10×)

Reagent	Quantity (for 250 mL)	Final concentration (10×)
NaCl	17 g	1.16 M
KCl	1 g	54 mM
$NaH_2PO_4 \cdot H_2O$	0.35 g	10 mM
Glucose	2.5 g	1%
Phenol red (0.5%)	2.5 mL	0.005%

Bring to 250 mL with ddH_2O and filter to sterilize.

Endothelial Cell Growth Medium

Reagent	Volume	Final concentration
Neurobasal medium (Gibco 21103-049)	18.9 mL	95%
SATO Supplement, NB-based (100×) <R>	200 µL	1×
T3 stock (4 µg/mL) <R>	200 µL	40 ng/mL
L-Glutamine (200 mM; Gibco 25030-081)	200 µL	2 mM
Sodium pyruvate (100 mM; Gibco 11360-070)	200 µL	1 mM
Penicillin/streptomycin (Gibco 15140-122)	200 µL	100 U/mL (penicillin) 100 µg/mL (streptomycin)
Insulin (0.5 mg/mL) <R>	200 µL	5 µg/mL
NAC stock (5 mg/mL) <R>	20 µL	5 µg/mL
Puromycin (10 mg/mL; Sigma P9620)[a]	2 µL	1 µg/mL
Forskolin stock (4.2 mg/mL)[a] <R>	20 µL	4.2 µg/mL
FCS (100%; heat-inactivated for 30 min at 55°C; Gibco 16000-036)[a]	100 µL	0.5%
Forskolin stock (4.2 mg/mL)[b] <R>	20 µL	4.2 µg/mL
bFGF (50 µg/mL)[b] <R>	20 µL	50 ng/mL
FCS (100%; heat-inactivated for 30 min at 55°C; Gibco 16000-036)[b]	100 µL	0.5%

Filter the medium through a prerinsed 0.22-µm filter into a conical tube. Store at 4°C.

[a]If the endothelial cells come from rodents older than P20, include these components in the medium on Days 1, 2, and 3 of culture to kill other types of cells.

[b]If the endothelial cells come from rodents older than P20, include these components in the medium beginning on Day 4 of culture to promote brain endothelial cell proliferation. If the cells come from rodents P20 and younger, include these components in the medium from Day 1 onward.

Ethanol-Washed Glass Coverslips

Extensively wash 12-mm glass coverslips (Carolina Biological Supply 633029) in 70% ethanol. Perform the washes on a platform shaker in a beaker, with enough motion to lightly agitate the coverslips but not break too many. Wash the coverslips for about 1 mo, exchanging the ethanol approximately every day. (It is fine to skip some exchanges.) Store the washed coverslips in 70% ethanol until use.

Forskolin Stock (4.2 mg/mL)

To prepare, add 1 mL of sterile DMSO to a 50-mg bottle of forskolin (Sigma-Aldrich F6886) and pipette up and down until the powder is fully resuspended. Transfer to a 15-mL conical tube and add an additional 11 mL of DMSO to achieve a final concentration of 4.2 mg/mL. Store in 20- and 80-µL aliquots at −20°C.

High-Ovomucoid Stock Solution (6×)

BSA (Sigma-Aldrich A8806)
D-PBS (Thermo Scientific HyClone SH3026401)
NaOH (1 N)
Trypsin inhibitor (Worthington LS003086)

1. Add 6 g of BSA to 150 mL D-PBS.
2. Add 6 g of trypsin inhibitor and mix to dissolve.

Cite this protocol as *Cold Spring Harb Protoc;* doi:10.1101/pdb.prot074963

3. Add at least 1.5 mL of 1 N NaOH to adjust the pH; continue adding NaOH as necessary to bring up the pH to 7.4.
4. Bring the volume to 200 mL with D-PBS.
5. Filter-sterilize through a 0.22-µm filter.
6. Make 1.0-mL aliquots and store at −20°C.

Inhibitor Stock Solution

Reagent	Volume	Final concentration
EBSS (10×) (Sigma-Aldrich E7510)	50 mL	1×
D$^+$-Glucose (30%)	6 mL	0.46%
$NaHCO_3$ (1 M)	13 mL	26 mM

Bring the volume to 500 mL with ddH_2O and filter-sterilize through a 0.22-µm filter.

Insulin (0.5 mg/mL)

1. Working under a sterile tissue culture hood, pre-rinse a 0.22-µm filter with sterile H_2O. Discard the flowthrough.
2. Add 50 mg of insulin (either recombinant human insulin [Sigma-Aldrich I2643] or bovine pancreas insulin [Sigma-Aldrich I6634]) to 100 mL of sterile H_2O in a 200-mL beaker. Immediately add 500 µL of 1 N HCl to adjust the pH.
3. Stir the mixture with a sterile pipette to dissolve the insulin completely; avoid creating bubbles.
4. Filter the solution through the prerinsed 0.22-µm filter.
5. Make 5-mL aliquots and store at 4°C for up to 4–6 wk. For use in media, dilute 100× to a concentration of 5 µg/mL.

Low-Ovomucoid Stock Solution (10×)

To prepare, add 3 g of BSA (Sigma-Aldrich A8806) to 150 mL D-PBS. Mix well. Add 3 g of trypsin inhibitor (Worthington LS003086) and mix to dissolve. Add ~1 mL of 1 N NaOH to adjust the pH to 7.4. Bring the volume to 200 mL with D-PBS. Filter-sterilize through a 0.22-µm filter. Make 1.0-mL aliquots and store at −20°C.

NAC Stock (5 mg/mL)

To prepare, dissolve 50 mg of *N*-acetyl-L-cysteine (NAC) powder (Sigma-Aldrich A8199) in 10 mL of Neurobasal Medium (Gibco/Life Technologies 21103). (The solution will be yellowish.) Filter through a 0.22-µm filter. Prepare 20- and 80-µL aliquots and store them at 4°C.

Papain Buffer

Reagent	Amount (for 250 mL)	Final concentration
EBSS stock (10×) <R>	25 mL	1×
$MgSO_4$ (100 mM)	2.5 mL	1 mM
Glucose (30%)	3 mL	0.46%
EGTA (0.5 M)	1 mL	2 mM
$NaHCO_3$ (1 M)	6.5 mL	26 mM

Bring volume up to 250 mL with ddH_2O and filter to sterilize.

Poly-D-Lysine (PDL) Stock (1 mg/mL)

1. Resuspend poly-D-lysine (PDL; Sigma-Aldrich P6407; molecular weight 70–150 kDa) at 1 mg/mL in borate buffer by combining 50 mg of PDL and 50 mL of 0.15 M boric acid (pH 8.4).
2. Filter to sterilize, then aliquot (e.g., 100 µL/tube). Store aliquots at −20°C.

SATO Supplement, NB-Based (100×)

1. Prepare the following stock solutions (these should be made fresh; do not reuse).
   - Combine 2.5 mg of progesterone (Sigma-Aldrich P8783) and 100 µL of ethanol to make a progesterone stock solution.
   - Combine 4.0 mg of sodium selenite (Sigma-Aldrich S5261), 10 µL of 1 N NaOH, and 10 mL of Neurobasal (NB, Gibco 21103-049) to make a sodium selenite stock solution.
2. Add the following to 80 mL of Neurobasal medium:

Reagent	Quantity	Final concentration in medium (1×)
BSA (Sigma-Aldrich A4161)	800 mg	100 µg/mL
Transferrin (Sigma-Aldrich T1147)	800 mg	100 µg/mL
Putrescine dihydrochloride (Sigma-Aldrich P5780)	128 mg	16 µg /mL
Progesterone stock solution	20 µL	60 ng/mL (0.2 µM)
Sodium selenite stock solution	800 µL	40 ng/mL

3. Mix well, and filter-sterilize through a prerinsed 0.22-µm filter. Make 200-µL or 800-µL aliquots, and store at −20°C.

T3 Stock (4 µg/mL)

1. Dissolve 4 µg of thyroid hormone (T3; Sigma-Aldrich T6397) in 500 µL of 1 N NaOH to prepare a solution of 0.8 µg/100 µL.
2. Add 75 µL of the T3 solution from Step 1 to 150 mL of Dulbecco's phosphate-buffered saline (D-PBS; Invitrogen 14287-080).
3. Filter solution through a filter-sterilization unit, discarding the first 10 mL.
4. Aliquot (e.g., 200 µL/tube) and then store at −20°C.

REFERENCE

Zhou L, Sohet F, Daneman R. 2014. Purification and culture of central nervous system endothelial cells. *Cold Spring Harb Protoc* doi: 10.1101/pdb.top070987.

CHAPTER 8

Purification and Culture of Oligodendrocyte Lineage Cells

Jason C. Dugas[1,2,4] and Ben Emery[1,3,4]

[1]*Stanford University School of Medicine, Department of Neurobiology, Stanford, California 94305-5125*

Oligodendrocytes are the cells of the vertebrate central nervous system responsible for forming myelin sheaths, which are essential for the rapid propagation of action potential. The formation of oligodendrocytes and myelin sheaths is tightly regulated, both temporally and spatially, by a number of extracellular and intracellular factors. For example, notch ligands, thyroid hormones, and mitogens such as platelet-derived growth factor (PDGF) and fibroblast growth factor (FGF) can all interact with oligodendrocyte precursor cell–expressed receptors to impact proliferation, differentiation, and myelin gene expression. To facilitate oligodendrocyte biology research, we have developed a technique using immunopanning to isolate different stages of the oligodendrocyte lineage, oligodendrocyte precursor cells and/or postmitotic oligodendrocytes, from postnatal rat or mouse brains. These cells can be cultured in defined, serum-free media in conditions that either promote differentiation into mature oligodendrocytes or continued proliferation as immature oligodendrocyte precursors. These cells represent an ideal system in which to study the regulation of oligodendrocyte proliferation, migration, differentiation, myelin gene expression, or other fundamental aspects of oligodendrocyte biology.

OLIGODENDROCYTE DEVELOPMENT

Oligodendrocytes Myelinate CNS Axons

In vertebrates, myelination has evolved as a method of electrically insulating axons to promote rapid, energy-efficient saltatory action potential propagation. In the central nervous system (CNS), this myelination is performed by oligodendrocytes (OLs), which extend large processes to extensively wrap selected nearby axons (Baumann and Pham-Dinh 2001). The importance of the development of normal CNS myelin is demonstrated in several diseases that result in CNS demyelination, such as Canavan's disease, Alexander's disease, Pelizaeus–Merzbacher disease, adrenoleukodystrophy, and multiple sclerosis, in which loss of myelin often results in severe neurological deficits and premature death in affected patients (Baumann and Pham-Dinh 2001).

Regulation of OL Differentiation and Myelination

During normal development, the committed precursors of OLs, oligodendrocyte precursor cells (OPCs), are generated from stereotypical germinal regions of the CNS neural tube (Rowitch 2004). OPCs then proliferate and migrate throughout the CNS before differentiating into OLs that myelinate nearby axons. OL differentiation and myelination is normally tightly regulated in the CNS, as the

[2]Present address: Myelin Repair Foundation, Saratoga, California 95070.

[3]Present address: Department of Anatomy and Neurobiology and Florey Institute of Neuroscience and Mental Health, University of Melbourne, Victoria 3010, Australia.

[4]Correspondence: jcdugas@alum.mit.edu; emeryb@unimelb.edu.au

Cite this introduction as *Cold Spring Harb Protoc*; doi:10.1101/pdb.top074898

regional progression of CNS myelination occurs in a very predictable manner in both humans and rodents (Brody et al. 1987; Rozeik and Von Keyserlingk 1987; Kinney et al. 1988; Hamano et al. 1996, 1998). The location and timing of CNS myelination is regulated in large part by controlling the progression of OL differentiation, as the expression of markers of OL maturity is rapidly followed by the initiation of myelination in the CNS (Baumann and Pham-Dinh 2001).

OL differentiation can be regulated in various ways. Intracellularly, several transcription factors have been shown to be required for the generation of mature OLs and myelin; some of the most well known of these genes include *Olig1*, *Olig2*, *Nkx2.2*, and *Sox10* (Qi et al. 2001; Zhou et al. 2001; Stolt et al. 2002; Zhou and Anderson 2002). Many of these genes are expressed in both immature OPCs and mature OLs, indicating that they are likely required for the specification of the OL lineage, and may not direct the transition from a proliferating OPC to a mature, postmitotic OL. More recently, we and others have identified additional transcriptional regulators that appear to be involved more specifically in terminal OL differentiation (Dugas et al. 2006; Sohn et al. 2006; Wang et al. 2006; Emery et al. 2009). In addition, genomic expression analyses performed in the Barres lab have indicated that OL differentiation proceeds in discernible temporal and likely functional stages (Dugas et al. 2006; Cahoy et al. 2008).

Extracellularly, a number of factors have been shown to influence OL differentiation. Locally, binding of OPC-expressed Notch receptors to membrane-bound Notch ligands (such as Jagged) can prevent OL differentiation (Wang et al. 1998; Givogri et al. 2002). More globally, OL differentiation can be influenced by signaling from several soluble extracellular factors, such as OPC mitogens and the thyroid hormone triiodothyronine (T3) (Temple and Raff 1986; Barres et al. 1993, 1994a; Baumann and Pham-Dinh 2001).

Mitogens are Required for OPC Proliferation

Several extrinsically derived mitogens have been shown to promote OPC proliferation and inhibit differentiation (Baumann and Pham-Dinh 2001). In culture, withdrawal of OPC mitogens from the culture medium results in rapid cessation of proliferation and initiation of OL differentiation (Barres et al. 1993, 1994a). Manipulation of mitogen levels in vivo has also been shown to affect normal OL differentiation: Mice lacking platelet-derived growth factor-AA (PDGF-AA) had severely reduced numbers of OPCs and, as a consequence, eventually reduced numbers of mature OLs (Fruttiger et al. 1999), whereas artificial overexpression of the OPC mitogen PDGF in vivo can transiently increase numbers of proliferating OPCs (Barres et al. 1992; Calver et al. 1998). Interestingly, these studies demonstrate that elevation of PDGF increases both OPC and OL numbers early in development, potentially indicating that other factors in addition to low mitogen levels could promote OL differentiation.

T3 Promotes Clock-Mediated OL Differentiation

Mitogens such as PDGF can maintain OPCs in an undifferentiated state in vitro, but only when cells are cultured in the absence of additional factors that are normally present in vivo, such as T3 or retinoic acid (RA) (Barres et al. 1994a; Tang et al. 2000). In the presence of T3 or RA, OPCs differentiate into OLs even when maintained in saturating amounts of mitogens (Temple and Raff 1986; Barres et al. 1994a), and T3 triggers differentiation in the presence of mitogens both in perinatal and adult OPCs (Shi et al. 1998). Interestingly, these factors generally do not induce immediate OL differentiation. Instead, OPCs cultured in the presence of both saturating amounts of mitogens and T3 can proliferate for several days before initiating differentiation in response to T3. This delayed differentiation appears to occur synchronously within all the members of a dividing OPC clone, and the timing of differentiation seems to depend more on the passage of time than on the number of cell divisions (Temple and Raff 1986; Gao et al. 1997). For these reasons, this mode of promoting OL differentiation has been termed "clock-mediated" differentiation. Interestingly, withdrawal of mitogens and presentation of T3 do not promote OL differentiation via identical pathways,

 Cite this introduction as *Cold Spring Harb Protoc*; doi:10.1101/pdb.top074898

as indicated by the distinct regulation of several cell cycle control genes in OPCs in response to these distinct differentiation-promoting stimuli (Tokumoto et al. 2001).

METHODS FOR THE PURIFICATION AND CULTURE OF OLIGODENDROCYTE LINEAGE CELLS

Methods for purifying and culturing OL lineage cells were established as long ago as 1980 and have included shaking methods (McCarthy and de Vellis 1980), immunopanning (Barres et al. 1992), FACS sorting (Aguirre and Gallo 2004) and specification from neural precursor cells or stem cells (Vitry et al. 1999; Najm et al. 2011). The immunopanning and culturing method introduced here for the purification of optic nerve OPCs was first described by Barres et al. (1992). The initial protocol has been subsequently modified to allow purification of cells from different CNS regions of either rats or mice, and to better promote the survival and either proliferation or differentiation, respectively, of cultured OPCs and OLs in defined, serum-free medium (Barres et al. 1993, 1994a,b; Wang et al. 2001; Dugas et al. 2006; Cahoy et al. 2008).

The current purification and culture system has several advantages. Immunopanning is capable of isolating OPCs or OLs to >99.5% purity, which is generally higher than the "shakeoff" method initially described by McCarthy and de Vellis (1980). Immunopanning is also, in general, gentler on cells than FACS sorting, another method commonly used to isolate OPCs and OLs (Aguirre and Gallo 2004). In addition, transgenic mice expressing green fluorescent protein from the *CNP1* promoter used to isolate OL-lineage cells label both OPCs and OLs, and the same lines can give different proportions of OPCs and OLs depending on the age of the animal (Sohn et al. 2006). In contrast, immunopanning allows the separation of OPCs from OLs at various developmental ages. This method also allows for the purification of OPCs from any CNS region that can be segregated by dissection, which enables the assessment of regional heterogeneity in OPCs; potential regional OPC heterogeneity has been indicated in previous reports (Ibarrola and Rodriguez-Pena 1997; Power et al. 2002).

We have provided two immunopanning protocols based on this method: Protocol 1: Purification of Oligodendrocyte Precursor Cells from Rat Cortices by Immunopanning (Dugas and Emery) and Protocol 2: Purification of Oligodendrocyte Lineage Cells from Mouse Cortices by Immunopanning (Emery and Dugas). The first describes isolation of OPCs from rat or mouse cortices after initial depletion of postmitotic OLs. The second provides the methods used in Cahoy et al. (2008) to purify OPCs, immature OLs, and mature OLs from mouse cortices. It should be noted that to generate large numbers of postmitotic OLs it is usually more efficient to isolate OPCs and expand them in vitro in the presence of PDGF before inducing differentiation than it is to acutely isolate the mature OLs. The former in vitro proliferation and differentiation approach allows for the generation of tens of millions of oligodendrocytes, whereas typically only a few million postmitotic OLs at most can be acutely isolated from postnatal brains. Nevertheless, for some experiments (e.g., assessing gene expression or recombination rates in vivo) the latter approach can still be useful.

The serum-free culture system employed in these protocols has several experimental advantages. Purified OPCs can be extensively cultured in defined, serum-free media that either promote their continued proliferation (in the presence of mitogens and absence of T3) or their differentiation (by adding T3, withdrawing mitogens, or both). All of the components of the media are known and can be manipulated to address experimental questions. The absence of serum in the media also avoids any potentially confounding effects of serum in increasing the transformation of OPCs into type-2 astrocytes rather than OLs (Temple and Raff 1985). We and others have observed that OPCs signaled to differentiate in culture express most, if not all, of the genes expressed by OLs in vivo (Baumann and Pham-Dinh 2001; Dugas et al. 2006), making this a very useful system in which to precisely study the regulation of various aspects of OL differentiation in isolation, independent of the influences of heterogeneous cell types. In addition, OPCs purified in this way can also be combined with purified retinal ganglion cell axons to study the development of CNS myelin in vitro. Finally, the aspects of T3 that promote "clock-mediated" OL differentiation, coupled with the fact that T3 coordinates the timing of differentiation events in various tissues and cell types, raise the interesting possibility of utilizing pure cultured OPCs to study the more general mechanisms that regulate the timing of mammalian cell differentiation.

These protocols have been optimized for use in both rats and mice, each species having distinct advantages and disadvantages. With rats, a greater number of OPCs can be isolated from a smaller number of animals. In addition, the OPCs generated from rats tend to be more robust when subjected to various manipulations, such as passaging or transfecting. Finally, unlike mouse cells, rat OPCs do not require an additional supplement of B-27 for survival. As commercially obtained B-27 contains small but physiologically relevant amounts of T3, the study of the effects of T3 in mouse OPCs necessitates custom formulation of B-27 which lacks T3 (such as NS21; Chen et al. 2008). The obvious advantage of using mice rather than rats is that it enables researchers to isolate and culture oligodendrocyte lineage cells from transgenic, knockout, or conditional knockout animals (e.g., Belachew et al. 2003; He et al. 2007; Emery et al. 2009; Ye et al. 2009; Zhao et al. 2010). These cells can be cultured in isolation or added to neuronal–glial cocultures to assess the capacity of different genotypes to myelinate, enabling independent modulation of genes in the glia and neurons (Watkins et al. 2008). In addition, for genomic analyses, several commercially available resources (Affymetrix, Agilent, etc.) are more extensive and better annotated for mice than for rats.

REFERENCES

Aguirre A, Gallo V. 2004. Postnatal neurogenesis and gliogenesis in the olfactory bulb from NG2-expressing progenitors of the subventricular zone. *J Neurosci* **24:** 10530–10541.

Barres BA, Hart IK, Coles HS, Burne JF, Voyvodic JT, Richardson WD, Raff MC. 1992. Cell death and control of cell survival in the oligodendrocyte lineage. *Cell* **70:** 31–46.

Barres BA, Schmid R, Sendnter M, Raff MC. 1993. Multiple extracellular signals are required for long-term oligodendrocyte survival. *Development* **118:** 283–295.

Barres BA, Lazar MA, Raff MC. 1994a. A novel role for thyroid hormone, glucocorticoids and retinoic acid in timing oligodendrocyte development. *Development* **120:** 1097–1108.

Barres BA, Raff MC, Gaese F, Bartke I, Dechant G, Barde YA. 1994b. A crucial role for neurotrophin-3 in oligodendrocyte development. *Nature* **367:** 371–375.

Baumann N, Pham-Dinh D. 2001. Biology of oligodendrocyte and myelin in the mammalian central nervous system. *Physiol Rev* **81:** 871–927.

Belachew S, Chittajallu R, Aguirre AA, Yuan X, Kirby M, Anderson S, Gallo V. 2003. Postnatal NG2 proteoglycan-expressing progenitor cells are intrinsically multipotent and generate functional neurons. *J Cell Biol* **161:** 169–186.

Brody BA, Kinney HC, Kloman AS, Gilles FH. 1987. Sequence of central nervous system myelination in human infancy. I. An autopsy study of myelination. *J Neuropathol Exp Neurol* **46:** 283–301.

Cahoy JD, Emery B, Kaushal A, Foo LC, Zamanian JL, Christopherson KS, Xing Y, Lubischer JL, Krieg PA, Krupenko SA, et al. 2008. A transcriptome database for astrocytes, neurons, and oligodendrocytes: A new resource for understanding brain development and function. *J Neurosci* **28:** 264–278.

Calver AR, Hall AC, Yu WP, Walsh FS, Heath JK, Betsholtz C, Richardson WD. 1998. Oligodendrocyte population dynamics and the role of PDGF in vivo. *Neuron* **20:** 869–882.

Chen Y, Stevens B, Chang J, Milbrandt J, Barres BA, Hell JW. 2008. NS21: Re-defined and modified supplement B27 for neuronal cultures. *J Neurosci Methods* **171:** 239–247.

Dugas J, Emery B. 2013. Purification of oligodendrocyte precursor cells from rat cortices by immunopanning. *Cold Spring Harb Protoc* doi: 10.1101/pdb.prot070862.

Dugas JC, Tai YC, Speed TP, Ngai J, Barres BA. 2006. Functional genomic analysis of oligodendrocyte differentiation. *J Neurosci* **26:** 10967–10983.

Emery B, Dugas J. 2013. Purification of oligodendrocyte lineage cells from mouse cortices by immunopanning. *Cold Spring Harb Protoc* doi: 10.1101/pdb.prot073973.

Emery B, Agalliu D, Cahoy JD, Watkins TA, Dugas JC, Mulinyawe SB, Ibrahim A, Ligon KL, Rowitch DH, Barres BA. 2009. Myelin gene regulatory factor is a critical transcriptional regulator required for CNS myelination. *Cell* **138:** 172–185.

Fruttiger M, Karlsson L, Hall AC, Abramsson A, Calver AR, Bostrom H, Willetts K, Bertold CH, Heath JK, Betsholtz C, et al. 1999. Defective oligodendrocyte development and severe hypomyelination in PDGF-a knockout mice. *Development* **126:** 457–467.

Gao FB, Durand B, Raff M. 1997. Oligodendrocyte precursor cells count time but not cell divisions before differentiation. *Curr Biol* **7:** 152–155.

Givogri MI, Costa RM, Schonmann V, Silva AJ, Campagnoni AT, Bongarzone ER. 2002. Central nervous system myelination in mice with deficient expression of Notch1 receptor. *J Neurosci Res* **67:** 309–320.

Hamano K, Iwasaki N, Takeya T, Takita H. 1996. A quantitative analysis of rat central nervous system myelination using the immunohistochemical method for MBP. *Brain Res* **93:** 18–22.

Hamano K, Takeya T, Iwasaki N, Nakayama J, Ohto T, Okada Y. 1998. A quantitative study of the progress of myelination in the rat central nervous system, using the immunohistochemical method for proteolipid protein. *Brain Res* **108:** 287–293.

He Y, Dupree J, Wang J, Sandoval J, Li J, Liu H, Shi Y, Nave KA, Casaccia-Bonnefil P. 2007. The transcription factor Yin Yang 1 is essential for oligodendrocyte progenitor differentiation. *Neuron* **55:** 217–230.

Ibarrola N, Rodriguez-Pena A. 1997. Hypothyroidism coordinately and transiently affects myelin protein gene expression in most rat brain regions during postnatal development. *Brain Res* **752:** 285–293.

Kinney HC, Brody BA, Kloman AS, Gilles FH. 1988. Sequence of central nervous system myelination in human infancy. II. Patterns of myelination in autopsied infants. *J Neuropathol Exp Neurol* **47:** 217–234.

McCarthy KD, de Vellis J. 1980. Preparation of separate astroglial and oligodendroglial cell cultures from rat cerebral tissue. *J Cell Biol* **85:** 890–902.

Najm FJ, Zaremba A, Caprariello AV, Nayak S, Freundt EC, Scacheri PC, Miller RH, Tesar PJ. 2011. Rapid and robust generation of functional oligodendrocyte progenitor cells from epiblast stem cells. *Nat Methods* **8:** 957–962.

Cite this introduction as *Cold Spring Harb Protoc*; doi:10.1101/pdb.top074898

Power J, Mayer-Proschel M, Smith J, Noble M. 2002. Oligodendrocyte precursor cells from different brain regions express divergent properties consistent with the differing time courses of myelination in these regions. *Dev Biol* **245:** 362–375.

Qi Y, Cai J, Wu Y, Wu R, Lee J, Fu H, Rao M, Sussel L, Rubenstein J, Qiu M. 2001. Control of oligodendrocyte differentiation by the Nkx2.2 homeodomain transcription factor. *Development* **128:** 2723–2733.

Rowitch DH. 2004. Glial specification in the vertebrate neural tube. *Nat Rev Neurosci* **5:** 409–419.

Rozeik C, Von Keyserlingk D. 1987. The sequence of myelination in the brainstem of the rat monitored by myelin basic protein immunohistochemistry. *Brain Res* **432:** 183–190.

Shi J, Marinovich A, Barres BA. 1998. Purification and characterization of adult oligodendrocyte precursor cells from the rat optic nerve. *J Neurosci* **18:** 4627–4636.

Sohn J, Natale J, Chew LJ, Belachew S, Cheng Y, Aguirre A, Lytle J, Nait-Oumesmar B, Kerninon C, Kanai-Azuma M, et al. 2006. Identification of Sox17 as a transcription factor that regulates oligodendrocyte development. *J Neurosci* **26:** 9722–9735.

Stolt CC, Rehberg S, Ader M, Lommes P, Riethmacher D, Schachner M, Bartsch U, Wegner M. 2002. Terminal differentiation of myelin-forming oligodendrocytes depends on the transcription factor Sox10. *Genes Dev* **16:** 165–170.

Tang DG, Tokumoto YM, Raff MC. 2000. Long-term culture of purified postnatal oligodendrocyte precursor cells. Evidence for an intrinsic maturation program that plays out over months. *J Cell Biol* **148:** 971–984.

Temple S, Raff MC. 1985. Differentiation of a bipotential glial progenitor cell in a single cell microculture. *Nature* **313:** 223–225.

Temple S, Raff MC. 1986. Clonal analysis of oligodendrocyte development in culture: Evidence for a developmental clock that counts cell divisions. *Cell* **44:** 773–779.

Tokumoto YM, Tang DG, Raff MC. 2001. Two molecularly distinct intracellular pathways to oligodendrocyte differentiation: Role of a p53 family protein. *EMBO J* **20:** 5261–5268.

Vitry S, Avellana-Adalid V, Hardy R, Lachapelle F, Baron-Van Evercooren A. 1999. Mouse oligospheres: From pre-progenitors to functional oligodendrocytes. *J Neurosci Res* **58:** 735–751.

Wang S, Sdrulla AD, diSibio G, Bush G, Nofziger D, Hicks C, Weinmaster G, Barres BA. 1998. Notch receptor activation inhibits oligodendrocyte differentiation. *Neuron* **21:** 63–75.

Wang S, Sdrulla A, Johnson JE, Yokota Y, Barres BA. 2001. A role for the helix-loop-helix protein Id2 in the control of oligodendrocyte development. *Neuron* **29:** 603–614.

Wang SZ, Dulin J, Wu H, Hurlock E, Lee SE, Jansson K, Lu QR. 2006. An oligodendrocyte-specific zinc-finger transcription regulator cooperates with Olig2 to promote oligodendrocyte differentiation. *Development* **133:** 3389–3398.

Watkins TA, Emery B, Mulinyawe S, Barres BA. 2008. Distinct stages of myelination regulated by γ-secretase and astrocytes in a rapidly myelinating CNS coculture system. *Neuron* **60:** 555–569.

Ye F, Chen Y, Hoang T, Montgomery RL, Zhao XH, Bu H, Hu T, Taketo MM, van Es JH, Clevers H, et al. 2009. HDAC1 and HDAC2 regulate oligodendrocyte differentiation by disrupting the β-catenin–TCF interaction. *Nat Neurosci* **12:** 829–838.

Zhao X, He X, Han X, Yu Y, Ye F, Chen Y, Hoang T, Xu X, Mi QS, Xin M, et al. 2010. MicroRNA-mediated control of oligodendrocyte differentiation. *Neuron* **65:** 612–626.

Zhou Q, Anderson DJ. 2002. The bHLH transcription factors OLIG2 and OLIG1 couple neuronal and glial subtype specification. *Cell* **109:** 61–73.

Zhou Q, Choi G, Anderson DJ. 2001. The bHLH transcription factor Olig2 promotes oligodendrocyte differentiation in collaboration with Nkx2.2. *Neuron* **31:** 791–807.

Protocol 1

Purification of Oligodendrocyte Precursor Cells from Rat Cortices by Immunopanning

Jason C. Dugas[1,2,4] and Ben Emery[3]

[1]*Myelin Repair Foundation, Saratoga, California 95070;* [2]*Department of Neurobiology, School of Medicine, Stanford University, Stanford, California 94305-5125;* [3]*Department of Anatomy and Neuroscience and the Florey Institute of Neuroscience and Mental Health, University of Melbourne, Melbourne, Victoria 3010, Australia*

This protocol describes how to purify oligodendrocyte precursor cells (OPCs) from postnatal rodent brains. The method utilizes an immunopanning technique to first remove unwanted cells by negative selection and then purify OPCs by positive selection and subsequent enzymatic release from the final panning plate. Included are modifications that allow for purification and culturing of OPCs from mouse instead of rat tissue and for use of optic nerves instead of whole brains. The method for isolating OPCs from whole brain can be used for isolating OPCs from any specific region of the brain, provided that the area can be dissected away from the rest of the tissue. Suggested culture media for maintaining proliferating OPCs or inducing oligodendrocyte (OL) differentiation are also described.

MATERIALS

It is essential that you consult the appropriate Material Safety Data Sheets and your institution's Environmental Health and Safety Office for proper handling of equipment and hazardous material used in this protocol.

RECIPES: Please see the end of this protocol for recipes indicated by <R>. Additional recipes can be found online at http://cshprotocols.cshlp.org/site/recipes.

Reagents

Antibodies

- Anti-mouse Thy 1.2 monoclonal (Serotec MCA02R) (for mouse OPCs only)
- Anti-O1 antibody (Millipore MAB344)
- Anti-O4 antibody (Millipore MAB345)
- Goat anti-mouse IgG + IgM (Jackson ImmunoResearch 115-005-044)
- Goat anti-mouse IgM, μ chain specific (Jackson ImmunoResearch 115-005-020)

BSA (4%)

To prepare a stock of 4% BSA in Dulbecco's phosphate-buffered saline (D-PBS), dissolve 8 g of BSA (Sigma-Aldrich A4161) in 150 mL of D-PBS (HyClone SH30264.01) at 37°C. Adjust the pH to 7.4 with ~1 mL of 1 N NaOH. Bring the volume to 200 mL. Filter through a 0.22-μm filter. Store in 1-mL aliquots at –20°C.

DMEM-SATO base growth medium <R>

DNase I

---

[4]Correspondence: jcdugas@me.com

Cite this protocol as *Cold Spring Harb Protoc*; doi:10.1101/pdb.prot070862

On ice, dissolve 12,500 U of DNase I (Worthington Biochemical LS002007) per 1 mL of chilled Earle's Balanced Salt Solution (EBSS; Invitrogen 14155-063). Filter-sterilize on ice. Aliquot (e.g., 200 µL/tube) and freeze overnight at –20°C. Store aliquots at –20 to –30°C.

Dulbecco's phosphate-buffered saline (D-PBS) with phenol red

Add 500 µL of 0.5% phenol red (Sigma-Aldrich P0290) per 500 mL bottle of Dulbecco's phosphate-buffered saline (D-PBS; Invitrogen 14287-080).

Dulbecco's phosphate-buffered saline (D-PBS) without Ca^{2+}/Mg^{2+} (Invitrogen 14190-144)

IMPORTANT: D-PBS without Ca^{2+}/Mg^{2+} is ONLY used for tissue dissection (see Steps 9 and 15). All other steps in the protocol refer to the standard D-PBS with phenol red. Cells will not stick to the panning plates in the Ca^{2+}/Mg^{2+}-free D-PBS.

Earle's balanced salt solution (EBSS) (Invitrogen 14155-063)

Ethanol (70%)

Ethanol-washed glass coverslips (optional) <R>

Fetal calf serum (FCS) (heat-inactivated)

Prepare 50-mL aliquots of fetal calf serum (Invitrogen 16000-036) in sterile conical tubes. Heat-inactivate aliquots for 30 min in a 55°C water bath, and then store at –20°C.

High-ovomucoid (high-ovo) stock solution (6×) <R>

Insulin stock (0.5 mg/mL) <R>

L-cysteine (Sigma-Aldrich C7477)

Low-ovomucoid (low-ovo) stock solution (10×) <R>

NaOH (1 M, sterile)

OPC culture medium <R>

Addition of PDGF and NT-3 to medium in the absence of T3 will promote OPC proliferation, whereas addition of T3 to medium in the absence of PDGF and NT-3 will promote rapid OL differentiation. Robust differentiation will also commence in medium lacking both mitogens (PDGF and NT-3) and T3, but OLs might be slightly less healthy. For the study of OPCs in the complete absence of T3-mediated signaling, prepare DMEM-SATO base growth medium without B-27. We have found that FGF is not required in the medium to inhibit differentiation and maintain OPCs as immature, proliferating cells.

Papain (Worthington Biochemical LS003126)

Papain buffer <R>

Phenol red solution (0.5%) (Sigma-Aldrich P0290)

Phosphate-buffered saline (PBS) (filter- or autoclave-sterilized)

Poly-D-lysine (PDL) stock (1 mg/mL) <R>

The 1 mg/mL solution is a 100× stock. Dilute it to 1× in sterile H_2O before use.

Ran-2 hybridoma cells (supernatant for rat OPC prep) (ATCC TIB-119)

Rat pups (Sprague–Dawley) (postnatal day 6–8)

One rat brain should yield ~2–3 × 10^6 OPCs, whereas one mouse brain should yield ~0.8–1 × 10^6 OPCs. Yields and initial viability of purified OPCs will progressively decline in correlation with the age of animals used: Older animals yield lower numbers of less healthy OPCs. Mouse pups at postnatal day 6–8 can also be used.

Tris-HCl (50 mM, pH 9.5) (filter- or autoclave-sterilized)

Trypsin stock

Dissolve trypsin (Sigma-Aldrich T9935) at 50,000 U/mL in Earle's Balanced Salt Solution (EBSS) (Invitrogen 14155-063) on ice. If the powder does not go into solution easily, stir with a magnetic stir bar or try altering the pH with NaOH or HCl as appropriate. When fully dissolved, aliquot (e.g., 200 µL/tube) and store at –80°C. For each new batch of trypsin stock, the optimal trypsinization time required to release OPCs from the final panning plate (Step 46) should be redetermined.

Trypan blue (Invitrogen 15250-061)

Equipment

Centrifuge (tabletop, with 15-mL/50-mL conical tube adaptors)

Conical tubes (15- and 50-mL, sterile)

Dissection microscope
Forceps (#5 or #55) (for optic nerve preparation)
Glass spreader (sterile)
Heat block base (with a flat insert set to 34°C in a sterile hood)
Hemocytometer slide (Hausser Scientific 3110)
Laminar flow tissue culture hood
Monoject syringes (sterile) (Fisher Scientific)
Needles (21- and 23-gauge) (for optic nerve preparation)
Nitex mesh filters

Nitex mesh 20-µm opening filters can be ordered from Amazon. Cut filters into approximately 3 inch × 3 inch squares. Wrap sets of 8–12 in aluminum foil and autoclave to sterilize.

Petri dish lid (6-cm, with a hole in the center to accommodate a 0.22-µm filter)

Create the hole using flamed forceps to melt an opening into the center of the lid, or use a drill.

Petri dishes (6-, 10-, and 15-cm) (Falcon or Nunc)
Scalpel blade handle
Scalpel blades (#10)
Scissors (large, for decapitation) (e.g., Roboz RS-6820)
Scissors (small, curved) (e.g., Roboz RS-5675)
Scissors (very fine) (for optic nerve preparation) (e.g., Roboz RS-5602)
Source of carbon dioxide (5% CO_2/95% O_2) with a line leading to a heat block in a sterile hood
Syringe filters (0.22-µm)
Tissue-culture incubator (set to 37°C, 10% CO_2)
Tissue-culture plates (plastic) (6- and 24-well) (Falcon or Nunc)
Water bath (set to 37°C)

METHOD

An overview of the method is provided in Figure 1. For any steps that require filter sterilization of reagents before use, prerinse the filter with 5–10 mL of base liquid (e.g., D-PBS) to remove any residual surfactants that may be detrimental to cell health. Discard the prefilter liquid before filtering the solution to be sterilized.

Preparation of Plates and Reagents

Perform Step 1 on the day before cell purification. Steps 2–5 can be performed the day before or the same day as cell purification. (PDL-coated coverslips and tissue culture plates are usable up to 6 d following preparation.) For users familiar with the protocol, Steps 2–5 also may be performed during the 90-min papain digestion (Step 17) or panning steps (Steps 35–39) to save time.

1. Coat the panning dishes with secondary antibodies as follows.

   i. Coat two 15-cm Petri dishes with 60 µL of goat anti-mouse IgG + IgM plus 20 mL of sterile 50 mM Tris-HCl (pH 9.5) per dish.

   ii. Coat one 10-cm Petri dish with 30 µL of goat anti-mouse IgM plus 10 mL of sterile 50 mM Tris-HCl (pH 9.5).

   iii. Swirl the plates until the surfaces are evenly coated with the antibody-Tris solution.

   iv. Incubate the panning plates overnight at 4°C.

   One set of panning plates and reagents is sufficient for one postnatal rat brain, one to three postnatal mouse brains, or one to three litters of rodent optic nerves.

2. Prepare the culture dishes as follows.

   i. Add 5 mL/1 mL/250 µL of 1× PDL per well to 10-cm/6-well/24-well tissue culture plates, respectively.

   ii. Swirl to evenly coat plastic.

Cite this protocol as *Cold Spring Harb Protoc*; doi:10.1101/pdb.prot070862

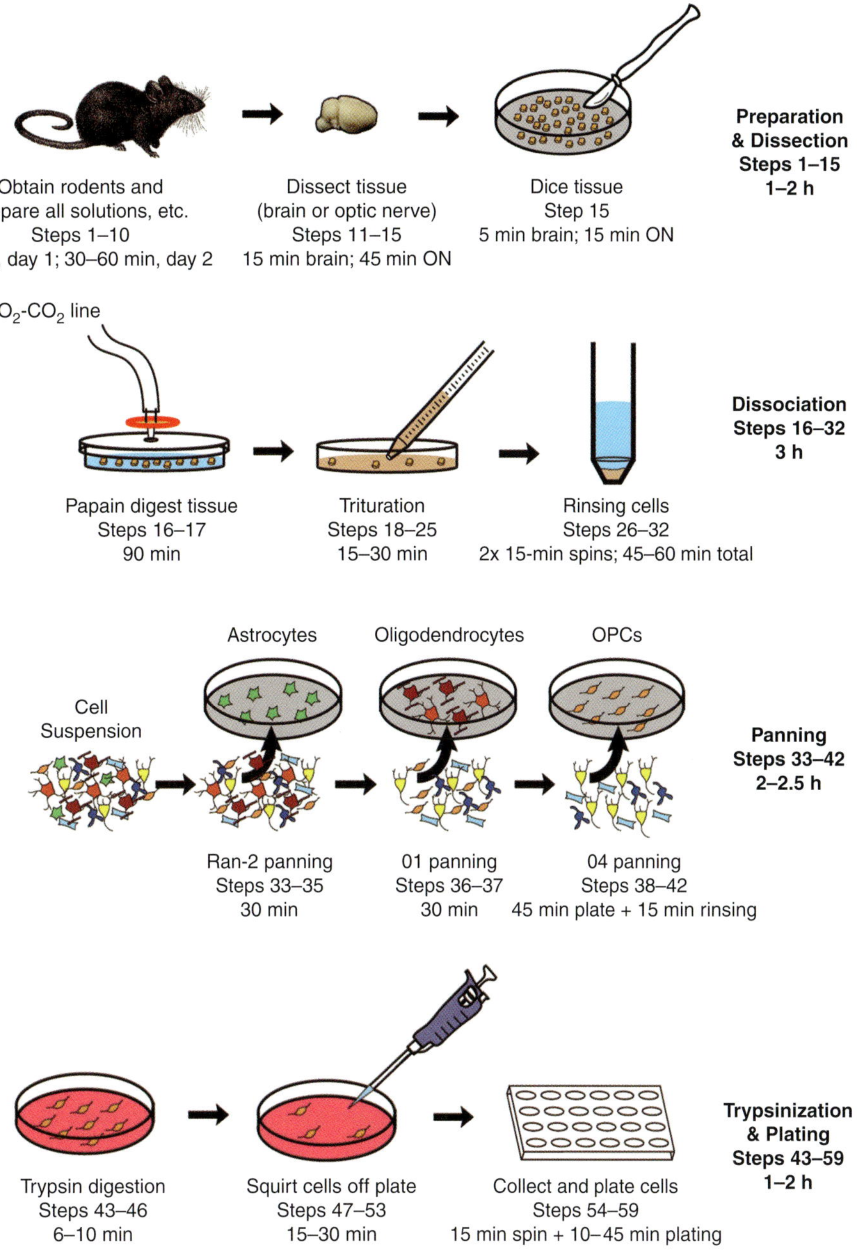

FIGURE 1. Immunopanning of OPCs. ON, optical nerves.

iii. Incubate for 20–60 min at room temperature.

iv. Rinse the plates three times with sterile H_2O.

v. Allow the plates to completely air-dry before use.

3. Prepare the coverslips as follows.

i. Rinse the ethanol-washed glass coverslips three times with sterile H_2O in a Petri dish.

ii. After the last rinse, suction away the remaining H_2O and separate the coverslips in the plate such that they are not touching each other or the sides of the plate.

iii. Allow the coverslips to completely air-dry.

This should take 5–10 min after suctioning.

iv. Carefully add 100 μL of 1× PDL to the center of each coverslip.

The solution should remain as a bubble on the coverslip.

v. Incubate for 30–60 min at room temperature.

vi. Rinse the coverslips three times with sterile H_2O.

vii. Transfer the coverslips to 24-well tissue culture plates, placing one coverslip per well.

viii. Suction any remaining H_2O from the wells, being careful to keep the coverslips centered and not touching the sides of the wells.

If the coverslips contact the sides of the wells, the preplating of OPCs will not work.

ix. Allow the coverslips to completely air-dry before use.

4. Prepare 40 mL of 0.2% BSA solution by adding 2 mL of 4% BSA stock to 38 mL of D-PBS. Store at 4°C.
5. Prepare the DMEM-SATO base medium (or the desired medium) for the cells.

Preparation for Cell Purification

Perform Steps 6–10 on the day of cell purification. For users familiar with the protocol, Step 10 also may be performed during the 90-min papain digestion (Step 17) to save time.

Do not allow the plates to dry at any stage.

6. Coat the panning dishes with primary antibodies as follows.

   i. Rinse each antibody-coated plate (prepared in Step 1) three times with PBS.

   ii. To prepare for use of rat cells, coat one 15-cm plate from Step 1 with 4 mL of Ran-2 hybridoma supernatant plus 8 mL of 0.2% BSA (prepared in Step 4).

   iii. To prepare for use of mouse cells, coat one 15-cm plate from Step 1 with 20 μL of Thy 1.2 antibody plus 12 mL of 0.2% BSA.

   iv. Coat one 15-cm plate from Step 1 with 12 μL (12 μg) of O1 antibody plus 12 mL of 0.2% BSA.

   v. Coat one 10-cm plate from Step 1 with 5 μL (5 μg) of O4 antibody plus 5 mL of 0.2% BSA.

   vi. Swirl to evenly coat plastic.

   vii. Incubate plates >2 h at room temperature.

7. Equilibrate 10 mL of papain buffer in a 6-mL Petri dish as follows.

   i. Presterilize a 6-cm dish lid with a central hole by spraying with 70% EtOH. Let air-dry.

   ii. Place the 6-cm dish containing papain buffer on a 34°C heat block in a sterile hood.

   iii. Place a 0.22-μm filter at the end of a 5% CO_2 / 95% O_2 line and place the sterile end of the filter through the hole in the sterilized lid.

   iv. Place the lid supplying gas over the papain buffer dish in the heat block and supply a gentle flow of gas.

8. Place 10 mL of EBSS containing 0.0005% phenol red into a 10% CO_2 incubator to equilibrate.
9. Add ~300 μL of D-PBS without Ca^{2+}/Mg^{2+} to a 6-cm Petri dish. Alternatively, if OPCs will be isolated from optic nerves rather than whole brain, use ~5 mL of D-PBS without Ca^{2+}/Mg^{2+}.
10. Prepare the solutions for dissociation, panning and trypsinization as follows.

   i. Prepare low-ovo working solution by adding 1 mL of 10× low-ovo stock to 9 mL of D-PBS.

   ii. Prepare high-ovo working solution by adding 1 mL of 6× high-ovo stock to 5 mL of D-PBS.

   iii. Prepare panning buffer by combining 1.5 mL of 0.2% BSA solution, 13.5 mL of D-PBS, and 150 μL of insulin stock (500 μg/mL).

Cite this protocol as *Cold Spring Harb Protoc*; doi:10.1101/pdb.prot070862

iv. Add 3 mL of heat-inactivated FCS to 7 mL of D-PBS and filter through a 0.22-µm filter to sterilize.

The phenol red color of the low- and high-ovo solutions should be orange (neutral pH) rather than red (too basic) or yellow (too acidic). If necessary, sterile 1 M NaOH can be used to equilibrate both solutions back to neutral pH.

Dissection

See online Movie 1 at cshprotocols.cshlp.org for an illustration of Steps 12–15.

11. Decapitate one Sprague–Dawley rat pup (or mouse pup) (P6-8) with sharp scissors.
12. Cut the skin along the top midline of the head, and peel back to reveal the skull.
13. Use curved scissors to open the skull: Insert the scissors at the opening at the back of the skull and cut along the left and right edge of the skull, above the ear to just above the eye socket.
14. Lift the skull bone to reveal the brain.
15. Dissect and harvest the brain or optic nerves as follows.

To Harvest the Brain

i. Cut away the olfactory bulbs at the front of the brain, then remove the rest of the brain.

The specific brain region of interest can also be dissected at this stage.

ii. Place the tissue in 300 µL of D-PBS without Ca^{2+}/Mg^{2+} in the dish prepared in Step 9.

iii. Dice the brain with a #10 scalpel into ~1-mm^3 chunks.

To Harvest the Optic Nerve

iv. After cutting the olfactory bulbs, lift the front of the brain up to reveal the optic nerves/optic tracts stretching from the base of the brain to the eye sockets. Cut the optic tracts immediately adjacent to the optic chiasm to leave the optic chiasm attached to the two optic nerves.

v. Cut at the base of each eyeball by inserting the curved scissors into the eye orbit behind the eyeball to detach the optic nerve from the eyeball.

vi. Grab the optic chiasm with tweezers and lift the optic chiasm plus the two optic nerves from the base of the skull.

vii. Place the tissue in the dish of 5 mL D-PBS without Ca^{2+}/Mg^{2+} prepared in Step 9.

viii. Dice the nerves with very small dissection scissors. Discard the optic chiasms.

Optic nerves should be harvested from approximately 10 pups to obtain ~500,000 OPCs.

Tissue Dissociation

16. Prepare the papain solution.

i. Move the equilibrated papain buffer prepared in Step 7 to a 15-mL conical tube and add 200 units of papain.

ii. Place this papain solution in a 37°C water bath for 5–15 min to allow the papain to dissolve.

iii. Weigh out and add 2 mg of L-cysteine to the solution.

iv. When fully dissolved, filter the papain solution through a 0.22-µm filter to sterilize.

v. Add 200 µL of DNase I stock solution to the papain solution.

17. Digest the tissue as follows.

Brain

i. Add the papain solution from Step 16 directly to the diced brain tissue in the 6-cm Petri dish.

ii. Place the dish of tissue in papain solution in an empty heat block at 34°C, and then cover with the same lid used to equilibrate the papain solution.

Brain dissociation digestions performed at 34°C slightly increase the health of the cells obtained compared to digestions at 37°C.

iii. Keep the tissue under 5% CO_2/95% O_2 gas flow at 34°C for 90 min, gently agitating the tissue every 15 min to ensure complete tissue digestion.

Optic Nerve

iv. Remove all of the D-PBS solution from the diced optic nerve tissue and add the papain solution.

v. Transfer the diced tissue to a sterile 50-mL conical tube and digest for 90 min in a 34–37°C water bath, gently agitating the tissue every 15 min.

18. Just before the digestion step is complete, add 100 µL of DNase I stock solution to the low-ovo working solution prepared in Step 10.
19. When the digestion is complete, transfer the tissue in papain solution to a sterile 15-mL conical tube and allow the tissue pieces to settle.
20. Carefully remove as much papain solution as possible from the tissue.
21. Gently add 2 mL of low-ovo solution with DNase I from Step 18 to the tissue to stop papain digestion. Allow the tissue to settle once again.
22. Remove and discard the low-ovo solution from the tissue.
23. Add 2 mL of fresh low-ovo solution with DNase I to the tissue.

    Dissociation is performed in low-ovo solution because the cells survive trituration better in a lower protein solution. High-ovo is subsequently used to fully quench any residual papain enzymatic activity.

24. Dissociate the tissue as follows.

Brain

See online Movie 2 at cshprotocols.cshlp.org for an illustration of Steps 24.i–24.v.

i. Gently pipette the tissue six to eight times through a 5-mL pipette.

ii. Let the tissue chunks settle for 1–2 min.

iii. Remove 1–1.5 mL of the resulting cell suspension, trying to avoid large tissue chunks. Place the collected cell suspension in a new, sterile 15-mL conical tube.

iv. Add 1 mL of low-ovo solution with DNase I and repeat Steps 24.i–24.iii.

v. Add 1 mL of low-ovo solution with DNase I, then follow Steps 24.i–24.iv, now triturating tissue through a 1-mL pipette tip; repeat until all of the tissue is dissociated and the low-ovo solution prepared in Step 18 is used up.

Optic Nerve

vi. Pipette the tissue six to eight times through a 1-mL pipette tip.

vii. Let the tissue chunks settle for 1–2 min.

viii. Remove 1–1.5 mL of the resulting cell suspension, trying to avoid large tissue chunks. Place the collected cell suspension in a new, sterile 15-mL conical tube.

ix. Add 1 mL of low-ovo solution with DNase I and repeat Steps 24.vi–24.viii.

x. Add 1 mL of low-ovo solution with DNase I, then follow Steps 24.vi–24.ix, now triturating the tissue three times through a 21-gauge needle followed by a 23-gauge needle, until the low-ovo solution prepared in Step 18 is used up.

25. (*Optional*) To determine cell viability and the effectiveness of the dissociation at this stage, remove a small aliquot (50–100 µL) of cell suspension for counting. Count as described in Step 55.

    See Troubleshooting.

Cite this protocol as *Cold Spring Harb Protoc*; doi:10.1101/pdb.prot070862

26. Pellet the cells at ~220*g* for 15 min in a tabletop centrifuge at room temperature.
27. Aspirate the supernatant, being careful not to disturb the cell pellet.
28. Resuspend the cell pellet in 6 mL of high-ovo solution prepared in Step 10.
29. Immediately pellet the cells at ~220*g* for 15 min in a tabletop centrifuge at room temperature.

    Leaving cells for an extended period of time in high-ovo solution will reduce cell viability.
30. Aspirate the supernatant, being careful not to disturb the cell pellet.
31. Resuspend the cell pellet in 6 mL of panning buffer prepared in Step 10.
32. Prewet a sterile Nitex mesh filter by filtering 2 mL of panning buffer into a sterile 50-mL tube. Filter the cells into the tube 1 mL at a time, and then rinse the filter with the remaining panning buffer.

Panning

Do not allow plates to dry out at any stage.

33. Immediately before panning dish use, rinse the panning dish three times with D-PBS.
34. Add the cell suspension from Step 32 to a rinsed Ran-2 (rat) or Thy1.2 (mouse) panning plate prepared in Step 6.
35. Incubate the plate for 30 min at room temperature, agitating at 15 min to ensure that all cells have an opportunity to adhere to the plate surface.
36. Gently shake the plate to loosen the nonadherent cells, and transfer the cell suspension to a rinsed O1 panning plate prepared in Step 6.
37. Incubate the plate for 30 min at room temperature, agitating for 15 min to ensure that all cells have an opportunity to adhere to the plate surface.
38. Gently shake the plate to loosen the nonadherent cells, and transfer the cell suspension to a rinsed O4 panning plate prepared in Step 6.
39. Incubate the plate for 45 min at room temperature, agitating every 15 min to ensure that all cells have an opportunity to adhere to the plate surface.
40. Shake the plate to loosen the nonadherent cells and pour off the cell suspension.
41. Rinse the plate six to eight times with D-PBS, shaking the plate at each rinse to loosen the nonadherent cells.

    OPCs adhere very strongly to the O4 panning plate, so shaking the final plate vigorously should not result in a significant loss of OPCs.
42. Before trypsinization, confirm visually under the microscope that nearly all nonadherent cells have been removed by rinsing. If needed, perform additional rinsing steps.

    There will always be a small number (<0.5%) of floating cells representing dislodged OPCs at the end of the rinsing steps.

    See Troubleshooting.

Trypsinization

See online Movie 3 at cshprotocols.cshlp.org for an illustration of Steps 48 and 49.

43. Once the OPCs are in their final D-PBS rinse, transfer 4 mL of the equilibrated EBSS prepared in Step 8 to a sterile tube and add 400 µL of trypsin stock solution.
44. Remove the D-PBS from the O4 plate and rinse the plate with the remaining 6 mL of equilibrated EBSS.
45. Pour off the EBSS and add the 4 mL of trypsin-EBSS solution prepared in Step 43.
46. Incubate the plate for 6–8 min in a 37°C incubator.

    Optimal trypsinization time should be determined with each batch of stock trypsin solution generated and should correspond to the time at which a large percentage of cells on the plate can be easily released by gentle pipetting. Trypsinization times that are either too short or too long can result in reduced cell viability.

47. Add 2 mL of the filtered 30% FCS solution prepared in Step 10 to the plate to stop trypsin digestion.
48. Dislodge the OPCs from the plate surface by squirting the FCS solution around the plate using a 1-mL pipette tip: Squirt the solution once around the entire circumference toward the center of the dish, once along the edge of the entire dish, and then once concentrating on center. Avoid scraping the tip of the pipette along the surface of the plate; instead squirt just above the surface. Avoid generating excess bubbles in the solution during squirting.
49. Remove the cell suspension from plate and place in a sterile 15-mL conical tube.
50. Place 5 mL of fresh 30% FCS in the plate and visualize the plate under the microscope to determine if there are particular regions of the plate that still contain adherent cells.

    Adherent cells are often found along the edges or in the exact center of the plate.
51. Repeat Steps 48 and 49 to dislodge and collect the remaining OPCs.
52. Rinse the plate with the remaining FCS and add the solution to the collection tube to collect the last remaining OPCs.
53. Remove a 50- to 100-μL aliquot of the cell suspension for counting.
54. Pellet the OPCs at ~220*g* for 15 min in a tabletop centrifuge at room temperature.
55. During Step 54, determine yield by adding an equal volume of trypan blue to the cell suspension aliquot from Step 53 and count the cells on a hemocytometer slide.

    Expected yields are as follows:
    - *One P6-8 rat brain, 2–2.5 × 10^6 OPCs*
    - *One P6-8 mouse brain, 0.8–1 × 10^6 OPCs*

    One litter (10 pups) P6-8 rat optic nerves, 250 × 10^3 OPCs

    See Troubleshooting.
56. After centrifugation, aspirate the supernatant and resuspend the cell pellet in a small volume of DMEM-SATO base growth medium.

Plating

OPCs are typically plated at a density of 10,000–20,000/coverslip in 24-well plates, 750,000–1,500,000/10-cm plate for immediate differentiation or short-term proliferation, or 1,000–5,000/50,000–200,000 for long term (5–7 d in vitro) proliferation.

57. Preplate the OPCs as follows.

    Coverslips

    i. Adjust the volume of the cell suspension in DMEM-SATO base growth medium to the number of desired cells per well/50 μL medium.

    ii. Place a 50-μL spot of cell suspension at the center of the individual coverslips prepared in Step 3.

    iii. Incubate for 20–45 min at 37°C to let the OPCs adhere to the coverslips.

    iv. Add the desired medium at 500 μL/well (see Step 58).

    Tissue culture plates

    v. Adjust the volume of the cell suspension in DMEM-SATO base growth medium to the number of desired cells per plate/300 μL medium.

    vi. Pipette 300 μL of cell suspension into a 10-cm PDL-coated tissue culture plate prepared in Step 2 and carefully spread the liquid with a sterile glass spreader, trying to avoid scraping the glass spreader on the plastic bottom of the plate.

    vii. Incubate for 7 min at 37°C to let the OPCs adhere to plate.

    viii. Add the desired medium at 10 mL/plate (see Step 58).

 Cite this protocol as *Cold Spring Harb Protoc*; doi:10.1101/pdb.prot070862

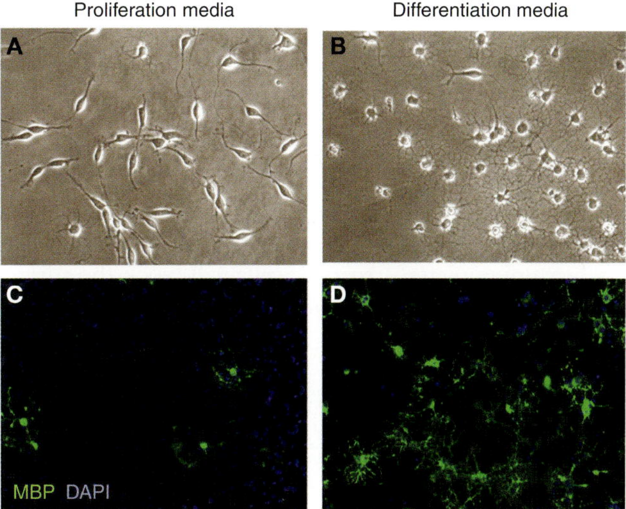

FIGURE 2. Purified OPCs cultured in proliferation- or differentiation-inducing media. OPCs purified from postnatal day-7 rats were cultured on glass coverslips for 4 d in defined, serum-free media that either contained saturating amount of mitogens (proliferation media [*A*,*C*]) or T3 with no mitogens (differentiation media [*B*,*D*]). Cells were imaged live on an inverted phase contrast microscope (*A*,*B*), or fixed, immunostained for expression of myelin basic protein (MBP, green), and mounted with DAPI to reveal nuclei (blue) (*C*,*D*).

ix. Alternatively, add the number of OPCs to be plated to 10 mL of the desired medium and add the solution directly to the PDL-coated 10-cm tissue culture plate.

This method results in slightly lower initial OPC viability.

58. Prepare OPC culture medium according to the cells used and the desired result.

PDGF, NT-3, and/or T3 may be added to the culture medium to promote OPC proliferation and/or differentiation (Fig. 2). We have found that, in general, withdrawal of mitogens to promote OL differentiation from OPCs leads to the rapid generation (3–4 d) of cultures that are predominantly (>90%) mature OLs, and 5%–10% type-2 astrocytes (by GFAP staining and morphology) if rat OPCs are used, and cultures that are 60%–70% mature OLs and 30%–40% type-2 astrocytes if mouse OPCs are used.

59. Incubate the OPC cultures at 37°C, 10% CO_2; replace 50% of the medium with fresh medium every 2–3 d.

To generate highly dense cultures of OPCs, we recommend using 20 ng/mL of PDGF (2× normal concentration) as the OPC cultures begin to get dense (> 25% confluency) to avoid unwanted differentiation. In addition, supplemental PDGF can be added directly to the cultures between feedings.

TROUBLESHOOTING

Problem (Steps 25 and 55): The final yield of OPCs is low.

Solution: Check the cell count of the total brain supernatant collected in Step 25. For whole brain, the yield at this step should be 20–30 × 10^6 total cells. If the yield is much lower, the dissociation steps (Steps 17–24) may have been the problem: Either the tissue was not sufficiently dissociated (if there are very few cells), or the trituration was too rough (if there is a very high percentage of trypan blue-positive dead cells). In addition, if the papain enzyme is bad, dissociation will be unsuccessful.

Problem (Steps 42 and 55): The final yield of OPCs is low and the cells were not very dense on the final planning plate at Step 42.

Solution: If the dissociation count in Step 25 is satisfactory, the next most likely source of the problem is the panning plates. If the cells were not very dense on the final panning plate at Step 42, there are several potential reasons.

- The cells were unhealthy and did not stick to the plate, possibly because of subtle problems in the trituration steps.
- The cells were rinsed too robustly. (Usually, however, the cells stick strongly to the O4 plate.)
- The panning buffer was made with D-PBS lacking Ca^{2+}/Mg^{2+}, or insulin was not added to the panning buffer. Either of these errors will lead to the cells not adhering strongly to the panning plates.
- The protocol was initiated using too much tissue and, thus, there was too much tissue in the panning plates. Excessive cells will block the surface of the plate and prevent the target cells (OPCs) from contacting and adhering to the plate surface.

Problem (Step 55): The final yield of OPCs is low, but the pretrypsinization density of OPCs on the final panning plate was high.
Solution: If the pretrypsinization density of OPCs was high at Step 42, then the trypsinization and final rinsing step might be a problem.

- If the trypsin solution is too weak, or if the trypsinization time is too short, it is very difficult to dislodge the OPCs from the final panning plate. In addition, cells that are dislodged will be unhealthy, leading to a low final yield. If after 10 min the OPCs are still stuck very strongly to the plate, try adding more trypsin to the plate or making a fresh batch of trypsin.
- Alternatively, greatly overtrypsinizing the OPCs will also be detrimental to cell health, so if the OPCs come off very easily with one squirt, try reducing the trypsinization time slightly.

RECIPES

CNTF Stock (10 μg/mL)

To prepare, dilute ciliary neurotrophic factor (CNTF; Peprotech 450-02) to 10 μg/mL with sterile 0.2% BSA in Dulbecco's phosphate-buffered saline (D-PBS; HyClone SH30264.01). Make 20-μL aliquots, flash freeze in liquid nitrogen, and store –80°C.

DMEM-SATO Base Growth Medium

1. Combine the following:

Reagent	Amount (for 20 mL)	Final concentration
Dulbecco's Modified Eagle's Medium (Invitrogen 11960-044)	19.5 mL	1×
SATO supplement (100×) <R>	200 μL	1×
Glutamine (200 mM; Invitrogen 25030-081)	200 μL	2 mM
Penicillin-streptomycin (Gibco/Life Technologies 15140-122)	200 μL	100 U/mL (penicillin)100 μg/mL (streptomycin)
Sodium pyruvate (100 mM; Invitrogen 11360-070)	200 μL	1 mM
Insulin stock (0.5 mg/mL) <R>	200 μL	5 μg/mL
N-Acetyl-L-cysteine (Sigma-Aldrich A8199) stock (5 mg/mL, prepared in DMEM)	20 μL	5 μg/mL
Trace Elements B (1000×; Cellgro 99-175-CI)	20 μL	1×
d-Biotin (Sigma-Aldrich B4639) stock (50 μg/mL)	4 μL	10 ng/mL

 Cite this protocol as *Cold Spring Harb Protoc;* doi:10.1101/pdb.prot070862

2. For mouse OPCs only, include 400 μL of 50× B-27 (Invitrogen 17504-044).

   Commercially available B-27 contains small but biologically relevant amounts of T3. If you wish to study OPCs in the absence of T3-mediated signaling, you should use an alternative custom supplement called NS21, omitting the T3 (Chen et al. 2008), or use rat OPCs, which do not require B-27 for survival.

3. Filter through a 0.22-μm filter to sterilize. Store at 4°C for up to 1 wk.

EBSS Stock (10×)

Reagent	Quantity (for 250 mL)	Final concentration (10×)
NaCl	17 g	1.16 M
KCl	1 g	54 mM
$NaH_2PO_4 \cdot H_2O$	0.35 g	10 mM
Glucose	2.5 g	1%
Phenol red (0.5%)	2.5 mL	0.005%

Bring to 250 mL with ddH_2O and filter to sterilize.

Ethanol-Washed Glass Coverslips

Extensively wash 12-mm glass coverslips (Carolina Biological Supply 633029) in 70% ethanol. Perform the washes on a platform shaker in a beaker, with enough motion to lightly agitate the coverslips but not break too many. Wash the coverslips for about 1 mo, exchanging the ethanol approximately every day. (It is fine to skip some exchanges.) Store the washed coverslips in 70% ethanol until use.

Forskolin Stock (4.2 mg/mL)

To prepare, add 1 mL of sterile DMSO to a 50-mg bottle of forskolin (Sigma-Aldrich F6886) and pipette up and down until the powder is fully resuspended. Transfer to a 15-mL conical tube and add an additional 11 mL of DMSO to achieve a final concentration of 4.2 mg/mL. Store in 20- and 80-μL aliquots at −20°C.

High-Ovomucoid Stock (6×)

1. Dissolve the following in 160–180 mL of Dulbecco's phosphate-buffered saline (D-PBS; Invitrogen 14287-080).

Reagent	Amount (for 200 mL)	Final concentration
BSA (Sigma-Aldrich A-8806)	6 g	30 mg/mL
Trypsin inhibitor (Worthington LS003086)	6 g	30 mg/mL

2. Adjust the pH to 7.4 with 10 N NaOH. Bring the volume to 200 mL with D-PBS, and then filter-sterilize.
3. Make 1-mL aliquots and store them at −20°C.

Insulin Stock (0.5 mg/mL)

To 20 mL of sterile water, add 10 mg of insulin (Sigma-Aldrich I6634) and 100 μL of 1.0 N HCl. Mix well. Filter through a 0.22-μm filter. Store at 4°C for 4–6 wk.

Low-Ovomucoid Stock Solution (10×)

To prepare, add 3 g of BSA (Sigma-Aldrich A8806) to 150 mL D-PBS. Mix well. Add 3 g of trypsin inhibitor (Worthington LS003086) and mix to dissolve. Add ~1 mL of 1 N NaOH to adjust the pH to 7.4. Bring the volume to 200 mL with D-PBS. Filter-sterilize through a 0.22-µm filter. Make 1.0-mL aliquots and store at −20°C.

NT-3 Stock (1 µg/mL)

1. Prepare a master stock of neurotrophin-3 (NT-3; Peprotech 450-03) at 1–0.1 mg/mL according to manufacturer's instructions. (Instructions may vary from lot to lot; generally, NT-3 is dissolved in buffer [e.g., Dulbecco's phosphate-buffered saline] plus 0.2% BSA.) Store at −80°C.
2. Prepare 0.2% BSA (Sigma-Aldrich A4161) in Dulbecco's phosphate-buffered saline (D-PBS; Invitrogen 14287-080). Filter-sterilize and chill.
3. Dilute an aliquot of NT-3 master stock to 1 µg/mL in sterile, chilled 0.2% BSA prepared in Step 2.
4. Aliquot the NT-3 working stock from Step 3 (e.g., 20 µL/tube) and snap-freeze in liquid nitrogen.
5. Store aliquots at −80°C.

OPC Culture Medium

1. To prepare full OPC culture medium, combine the following:
   20 mL DMEM-SATO base growth medium <R>
   20 µL Forskolin stock (4.2 mg/mL) <R>
   20 µL CNTF stock (10 µg/mL) <R>
2. Add growth factors as appropriate.
   - To promote OPC proliferation, add:
     20 µL PDGF stock (10 µg/mL) <R>
     20 µL NT-3 stock (1 µg/mL) <R>
   - To promote OL differentiation, omit PDGF and add:
     200 µL T3 stock (4 µg/mL) <R>
3. Store at 4°C for up to 3 d.

Papain Buffer

Reagent	Amount (for 250 mL)	Final concentration
EBSS stock (10×) <R>	25 mL	1×
$MgSO_4$ (100 mM)	2.5 mL	1 mM
Glucose (30%)	3 mL	0.46 %
EGTA (0.5 M)	1 mL	2 mM
$NaHCO_3$ (1 M)	6.5 mL	26 mM

Bring volume up to 250 mL with ddH_2O and filter to sterilize.

PDGF Stock (10 µg/mL)

1. Prepare a platelet-derived growth factor (PDGF) master stock by diluting PDGF (Peprotech 100-13A) at 1–0.1 mg/mL according to manufacturer's instructions. (Instructions may vary from lot to lot; generally, PDGF is dissolved in buffer [e.g., Dulbecco's phosphate-buffered saline] plus 0.2% BSA.) Store at −80°C.

Cite this protocol as *Cold Spring Harb Protoc*; doi:10.1101/pdb.prot070862

2. Prepare 0.2% BSA (Sigma-Aldrich A4161) in Dulbecco's phosphate-buffered saline (D-PBS; Invitrogen 14287-080). Filter-sterilize and chill.
3. Dilute an aliquot of PDGF master stock to 10 µg/mL in sterile, chilled 0.2% BSA prepared in Step 2.
4. Aliquot the PDGF working stock from Step 3 (e.g., 20 µL/tube) and snap-freeze in liquid nitrogen.
5. Store aliquots at −80°C.

Poly-D-Lysine (PDL) Stock (1 mg/mL)

1. Resuspend poly-D-lysine (PDL; Sigma-Aldrich P6407; molecular weight 70–150 kDa) at 1 mg/mL in borate buffer by combining 50 mg of PDL and 50 mL of 0.15 M boric acid (pH 8.4).
2. Filter to sterilize, and then aliquot (e.g., 100 µL/tube). Store aliquots at −20°C.

SATO Supplement (100×)

1. Prepare the following stock solutions (these should be made fresh; do not reuse):
   - Combine 5 mg of progesterone (Sigma-Aldrich P8783) and 200 µL of ethanol to make a progesterone stock solution.
   - Combine 4 mg of sodium selenite (Sigma-Aldrich S5261), 10 µL of 1 N NaOH, and 10 mL of Dulbecco's modified Eagle's medium (DMEM; Gibco/Life Technologies 11960-044) to make a sodium selenite stock solution.
2. Combine the following:

Reagent	Quantity (for 200 mL)	Final concentration (100×)
BSA (Sigma-Aldrich A4161)	2 g	10 mg/mL
Transferrin (Sigma-Aldrich T1147)	2 g	10 mg/mL
Putrescine (Sigma-Aldrich P5780)	320 mg	1.6 mg/mL
Progesterone stock solution	50 µL	6 µg/mL
Sodium selenite stock solution	2 mL	4 µg/mL

3. Bring to a total volume of 200 mL in DMEM, and then filter-sterilize. Aliquot and store at –20°C.

T3 Stock (4 µg/mL)

1. Dissolve 4 µg of thyroid hormone (T3; Sigma-Aldrich T6397) in 500 µL of 1 N NaOH to prepare a solution of 0.8 µg/100 µL.
2. Add 75 µL of the T3 solution from Step 1 to 150 mL of Dulbecco's phosphate-buffered saline (D-PBS; Invitrogen 14287-080).
3. Filter solution through a filter-sterilization unit, discarding the first 10 mL.
4. Aliquot (e.g., 200 µL/tube) and then store at −20°C.

REFERENCE

Chen Y, Stevens B, Chang J, Milbrandt J, Barres BA, Hell JW. 2008. NS21: Re-defined and modified supplement B27 for neuronal cultures. *J Neurosci Methods* **171:** 239–247.

Protocol 2

Purification of Oligodendrocyte Lineage Cells from Mouse Cortices by Immunopanning

Ben Emery[1,4] and Jason C. Dugas[2,3]

[1]*Department of Anatomy and Neuroscience and the Florey Institute of Neuroscience and Mental Health, University of Melbourne, Melbourne, Victoria 3010, Australia;* [2]*Myelin Repair Foundation, Saratoga, California 95070;* [3]*Department of Neurobiology, School of Medicine, Stanford University, Stanford, California 94305-5125*

Oligodendrocytes are the myelinating cells of the vertebrate central nervous system, responsible for generating the myelin sheath necessary for saltatory conduction. The use of increasingly sophisticated genetic tools, particularly in mice, has vastly increased our understanding of the molecular mechanisms that regulate development of the oligodendrocyte lineage. This increased reliance on the mouse as a genetic model has led to a need for the development of culture methods to allow the use of mouse cells in vitro as well as in vivo. Here, we present a protocol for the isolation of different stages of the oligodendrocyte lineage, oligodendrocyte precursor cells (OPCs) and/or postmitotic oligodendrocytes, from the postnatal mouse cortex using immunopanning. This protocol allows for the subsequent culture or biochemical analysis of these cells.

MATERIALS

It is essential that you consult the appropriate Material Safety Data Sheets and your institution's Environmental Health and Safety Office for proper handling of equipment and hazardous material used in this protocol.

RECIPES: Please see the end of this protocol for recipes indicated by <R>. Additional recipes can be found online at http://cshprotocols.cshlp.org/site/recipes.

Reagents

Antibodies

- Anti-PDGFRa antibody (e.g., rat anti-mouse CD140A; BD Pharmingen 558774)
- Goat anti-mouse IgG + IgM (Jackson ImmunoResearch 115-005-044)
- Goat anti-rat IgG (Jackson ImmunoResearch 112-005-167)

BSA stock (4%)

To prepare a stock of 4% BSA in Dulbecco's phosphate-buffered saline (D-PBS), dissolve 8 g of BSA (Sigma-Aldrich A4161) in 150 mL of D-PBS (HyClone SH30264.01) at 37°C. Adjust the pH to 7.4 with ~1 mL of 1 N NaOH. Bring the volume to 200 mL. Filter through a 0.22-µm filter. Store in 1-mL aliquots at –20°C.

BSL1 (Vector Laboratories L1100)

[4]Correspondence: emeryb@unimelb.edu.au

Cite this protocol as *Cold Spring Harb Protoc;* doi:10.1101/pdb.prot073973

DMEM-SATO base growth medium <R>

DNase I

On ice, dissolve 12,500 U of DNase I (Worthington Biochemical LS002007) per 1 mL of chilled Earle's Balanced Salt Solution (EBSS; Invitrogen 14155-063). Filter-sterilize on ice. Aliquot (e.g., 200 µL/tube) and freeze overnight at –20°C. Store aliquots at –20°C to –30°C.

Dulbecco's phosphate-buffered saline (D-PBS) with phenol red

Add 500 µL of 0.5% phenol red (Sigma-Aldrich P0290) per 500 mL bottle of Dulbecco's phosphate-buffered saline (D-PBS; Invitrogen 14287-080).

Dulbecco's phosphate-buffered saline (D-PBS) without Ca^{2+}/Mg^{2+} (Invitrogen 14190-144)

IMPORTANT: D-PBS without Ca^{2+}/Mg^{2+} is ONLY used at the tissue dissection step (see Steps 10 and 17). All other steps in the protocol refer to the standard D-PBS with phenol red. Cells will not stick to the panning plates in the Ca^{2+}-/Mg^{2+}-free D-PBS.

Earle's balanced salt solution (EBSS) (Invitrogen 14155-063)

Ethanol (70%)

Ethanol-washed glass coverslips (optional) <R>

Prepare ~1 mo in advance.

Fetal calf serum (FCS) (heat-inactivated)

Prepare 50-mL aliquots of fetal calf serum (Invitrogen 16000-036) in sterile conical tubes. Heat-inactivate aliquots for 30 min in a 55°C water bath, then store at –20°C.

High-ovomucoid (high-ovo) stock (6×) <R>

Hybridoma supernatants

- Anti-galactocerebroside (GalC) hybridoma supernatant (for all postmitotic [GalC$^+$] oligodendrocytes)
- Anti-MOG hybridoma supernatant (for mature [MOG$^+$] oligodendrocytes)

Insulin stock (0.5 mg/mL) <R>

L-cysteine (Sigma-Aldrich C-7477)

Low-ovomucoid (low-ovo) stock solution (10×) <R>

Mouse pups

One P7 mouse brain should yield about 0.5×10^6 OPCs; this yield will decline as the age of the animal increases. If purifying postmitotic oligodendrocytes, the yield is best at ~P12–P16. Beyond this age, dissociation of the tissue becomes progressively more difficult and the yield and viability of both OPCs and postmitotic oligodendrocytes are reduced.

If more than three pups are to be used, then the solutions and panning dish numbers will need to be increased. If several brains are to be processed in parallel (e.g., to compare two genotypes), adjust the number of panning plates and volumes of solutions accordingly.

OPC culture medium <R>

Addition of PDGF and NT-3 to medium in the absence of T3 will promote OPC proliferation, whereas addition of T3 to medium in the absence of PDGF and NT-3 will promote rapid OL differentiation. Robust differentiation will also commence in medium lacking both mitogens (PDGF and NT-3) and T3, but viability will be poor.

Papain (Worthington Biochemical LS003126)

Papain buffer <R>

Phenol red solution (0.5%) (Sigma-Aldrich P0290)

Phosphate-buffered saline (PBS) (filter- or autoclave-sterilized)

Poly-D-lysine (PDL) stock (1 mg/mL) <R>

The 1 mg/mL solution is a 100× stock. Dilute it to 1× in sterile H_2O before use.

Tris-HCl (50 mM, pH 9.5) (filter- or autoclave-sterilized)

Trypan blue (Invitrogen 15250-061)

Trypsin stock <R>

For each new batch of trypsin stock, the optimal trypsinization time required to release cells from the final panning plate (Step 52) should be redetermined.

Equipment

5% CO_2/95% O_2 source (with a line leading to a heat block in a sterile hood)
Centrifuge (tabletop, with 15-mL/50-mL conical tube adaptors)
Conical tubes (15- and 50-mL, sterile)
Flasks (175-cm^2) (optional)
Forceps (#5 or #55)
Heat block base (with a flat insert set to 34°C in a sterile hood)
Hemocytometer slide (Hausser Scientific 3110)
Laminar flow tissue culture hood
Monoject syringes (sterile) (Fisher Scientific)
Nitex mesh filters

Nitex mesh 20-µm opening filters can be ordered from Amazon. Cut filters into approximately 3 inch × 3 inch squares. Wrap sets of 8 to 12 in aluminum foil and autoclave to sterilize.

Petri dish lid (6-cm, with a hole in the center to accommodate a 0.22-µm filter)

Create the hole using flamed forceps to melt an opening into the center of the lid, or use a drill.

Petri dishes (6-, 10-, and 15-cm) (Falcon or Nunc)
Tissue-culture incubator (set to 37°C, 10% CO_2)
Scalpel blade handle
Scalpel blades (#10)
Scissors (large, for decapitation) (e.g., Roboz RS-6820)
Scissors (small, curved) (e.g., Roboz RS-5675)
Syringe filters (0.22-µm)
Tissue-culture plates (plastic) (6- or 24-well) (Falcon or Nunc) (optional)
Water bath (set to 37°C)

METHOD

An overview of the method is provided in Figure 1. For any steps that require filter sterilization of reagents before use, prerinse the filter with 5–10 mL of base liquid (e.g., D-PBS) to remove any residual surfactants that may be detrimental to cell health. Discard the prefilter liquid before filtering the solution to be sterilized.

Preparation of Plates and Reagents

Perform Step 1 on the day before cell purification. Steps 2–5 can be performed the day before or the same day as cell purification. (PDL-coated coverslips and tissue culture plates are usable up to 6 d following preparation.) For users familiar with the protocol, Steps 2–5 also may be performed during the 90-min papain digestion (Step 21) or panning steps (Steps 39–43) to save time.

1. Coat the panning dishes with secondary antibodies as follows.

   i. For OPCs (PDGFRa$^+$), coat a 10-cm Petri dish with 30 µL of goat anti-rat IgG plus 10 mL of sterile 50 mM Tris-HCl (pH 9.5).

   ii. For mature (MOG$^+$) oligodendrocytes, coat a 10-cm Petri dish with 30 µL of goat anti-mouse IgG + IgM plus 10 mL of sterile 50 mM Tris-HCl (pH 9.5).

   iii. For all postmitotic (GalC$^+$) oligodendrocytes, coat a 10-cm Petri dish with 30 µL of goat anti-mouse IgG + IgM plus 10 mL of sterile 50 mM Tris-HCl (pH 9.5).

   iv. Swirl the plates until the surfaces are evenly coated with the antibody-Tris solution.

   v. Incubate the panning plates overnight at 4°C.

   One set of panning plates and reagents is sufficient for one to three postnatal mouse brains.

2. If the isolated cells will be cultured at the end of the protocol, prepare the culture dishes as follows.

   i. Add 15 mL / 5 mL / 1 mL / 250 µL of 1× PDL per well to a 175-cm^2 flask / 10-cm dish / six-well plate / 24-well plate, respectively.

Cite this protocol as *Cold Spring Harb Protoc*; doi:10.1101/pdb.prot073973

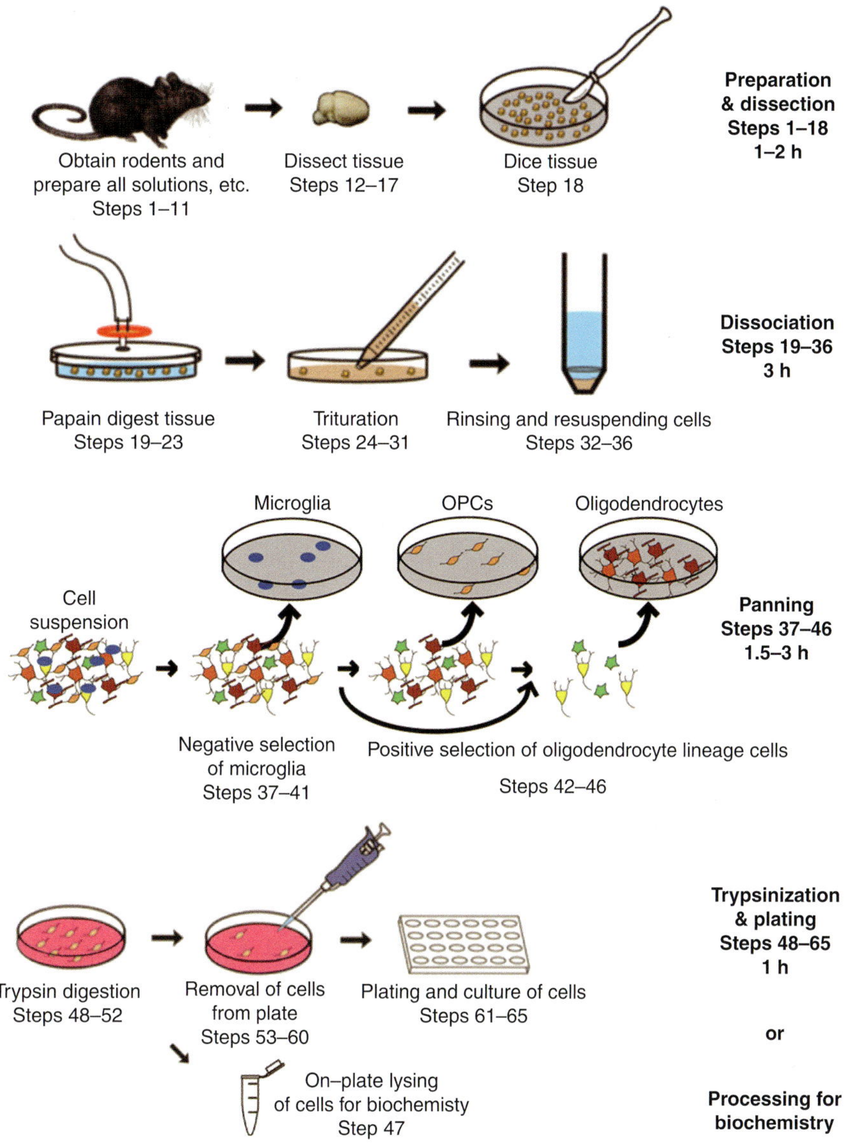

FIGURE 1. Purification of oligodendrocyte lineage cells by immunopanning.

ii. Swirl to evenly coat plastic.

iii. Incubate for 20–60 min at room temperature or at 4°C overnight.

iv. Rinse plates three times with sterile H_2O.

v. Allow plates to completely air-dry before use.

3. If the isolated cells will be cultured at the end of the protocol, prepare the coverslips as follows.

i. Rinse the ethanol-washed glass coverslips three times with sterile H_2O in a Petri dish.

ii. After the last rinse, suction away the remaining H_2O and separate the coverslips in the plate such that they are not touching each other or the sides of the plate.

iii. Allow the coverslips to completely air-dry.

This should take 5–10 min after suctioning.

iv. Carefully add 100 µL of 1× PDL to the center of each coverslip.

The PDL solution should remain as a bubble on the coverslip.

v. Incubate for 30–60 min at room temperature or at 4°C overnight.

vi. Rinse the coverslips three times with sterile H_2O.

vii. Transfer the coverslips to 24-well tissue culture plates, placing one coverslip per well.

viii. Suction any remaining H_2O from the wells, being careful to keep the coverslips centered and not touching the sides of the wells.

If the coverslips contact the sides of the wells, the preplating of the cells will not work.

ix. Allow the coverslips to completely air-dry before use.

4. Prepare 40 mL of 0.2% BSA solution by adding 2 mL of 4% BSA stock to 38 mL of D-PBS. Store at 4°C.

5. Prepare the DMEM-SATO base medium (or the desired medium) for the cells.

Preparation for Cell Purification

Perform Steps 6–11 on the day of cell purification. For users familiar with the protocol, Step 11 also may be performed during the 90-min papain digestion (Step 21) to save time.

Do not allow the plates to dry at any stage.

6. Coat two 15-cm dishes with BSL1 as follows.

   i. Dilute 40 µL of BSL1 in 40 mL of D-PBS.

   ii. Add 20 mL per dish and swirl to cover the entire surface.

7. Coat the panning dishes with primary antibodies as follows.

   i. Rinse each antibody-coated dish (prepared in Step 1) three times with PBS.

   ii. For anti-PDGFRa, add 12 mL of 0.2% BSA in D-PBS (prepared in Step 4) and 40 µL of rat anti-PDGFRa to a goat anti-rat IgG-coated dish from Step 1 and let sit at room temperature until use.

   iii. For anti-GalC, add 8 mL of 0.2% BSA in D-PBS and 4 mL of mouse anti-GalC supernatant to a goat anti-mouse IgG + IgM-coated dish from Step 1 and let sit at room temperature until use.

   iv. For anti-MOG, add 8 mL of 0.2% BSA in D-PBS and 4 mL mouse anti-MOG supernatant to a goat anti-mouse IgG + IgM-coated dish from Step 1 and let sit at room temperature until use.

   v. Swirl to evenly coat plastic.

   vi. Incubate for > 2 h at room temperature.

8. Equilibrate 10 mL of papain buffer in a 6-mL Petri dish as follows.

   i. Presterilize a 6-cm dish lid with a central hole by spraying with 70% ethanol. Let air-dry.

   ii. Place the 6-cm dish containing papain buffer on a 34°C heat block in a sterile hood.

   iii. Place a 0.22-µm filter at the end of a 5% CO_2 / 95% O_2 line and place the sterile end of the filter through the hole in the sterilized lid.

   iv. Place the lid supplying gas over the papain buffer dish in the heat block and supply a gentle flow of gas.

9. Place 10 mL of EBSS containing 0.0005% phenol red into a 10% CO_2 incubator to equilibrate.

10. Add ~300 µL of D-PBS without Ca^{2+}/Mg^{2+} to a 6-cm Petri dish (or, alternatively, each well of a six-well plate).

    If several brains are to be processed in parallel (e.g., to derive cells from several pups of different genotypes), Steps 17–21 can be carried out in a six-well plate, with one brain per well and a single hole in the lid for gas flow.

Cite this protocol as *Cold Spring Harb Protoc*; doi:10.1101/pdb.prot073973

11. Prepare the solutions for dissociation, panning and trypsinization as follows.
    i. Prepare low-ovo working solution by adding 1 mL of 10× low-ovo stock to 9 mL of D-PBS.
    ii. Prepare high-ovo working solution by adding 1 mL of 6× high-ovo stock to 5 mL of D-PBS.
    iii. Prepare panning buffer by combining 1.5 mL of 0.2% BSA solution, 13.5 mL of D-PBS, and 150 μL of insulin stock (500 μg/mL).
    iv. Add 3 mL of heat-inactivated FCS to 7 mL of D-PBS and filter through a 0.22-μm filter to sterilize.

    The phenol red color of the low- and high-ovo solutions should be orange (neutral pH) rather than red (too basic) or yellow (too acidic). If necessary, sterile 1 M *NaOH can be used to equilibrate both solutions back to neutral pH.*

Dissection

See online Movie 1 at cshprotocols.cshlp.org for an illustration of Steps 13–18.

12. Decapitate the mouse pup(s) with sharp scissors.
13. Cut the skin along the top midline of the head, and peel back to reveal the skull.
14. Use curved scissors to open the skull: Insert the scissors at the opening at the back of the skull and cut along the left and right edge of the skull, above the ear to just above the eye socket.
15. Lift the skull bone to reveal the brain.
16. Cut away the olfactory bulbs at the front of the brain, and then remove the rest of the brain.

    The specific brain region of interest can also be dissected at this stage.

17. Place the brain in 300 μL of D-PBS without Ca^{2+}/Mg^{2+} in the dish prepared in Step 10.
18. Dice the tissue with a #10 scalpel into ~1-mm^3 chunks.

Tissue Dissociation

19. Prepare the papain solution.
    i. Move the equilibrated papain buffer prepared in Step 8 to a 15-mL conical tube and add 200 units of papain.
    ii. Place this papain solution in a 37°C water bath for 5–15 min to allow the papain to dissolve.
    iii. Weigh out and add 2 mg of L-cysteine to the solution.
    iv. When fully dissolved, filter the papain solution through a 0.22-μm filter to sterilize.
    v. Add 200 μL of DNase I stock solution to the papain solution.
20. Add the papain solution from Step 19 directly to the diced brain tissue in the 6-cm Petri dish. Place the dish in an empty heat block at 34°C, and then cover with the same lid used to equilibrate the papain solution.
21. Keep the tissue under 5% CO_2/95% O_2 gas flow at 34°C for 90 min, gently agitating the tissue every 15 min to ensure complete tissue digestion.

    Brain dissociation digestions performed at 34°C slightly increase the health of the cells obtained compared to digestions at 37°C.

22. Just before the digestion step is complete, add 100 μL of DNase I stock solution to the low-ovo working solution prepared in Step 11.
23. When the digestion is complete, transfer the tissue in papain solution to a sterile 15-mL conical tube and allow the tissue pieces to settle.
24. Carefully remove as much papain solution as possible from the tissue.
25. Gently add 2 mL of low-ovo solution with DNase I from Step 22 to the tissue to stop papain digestion. Allow the tissue to settle once again.

26. Remove and discard the low-ovo solution from the tissue.
27. Add 2 mL of fresh low-ovo solution with DNase I to the tissue.

    Dissociation is performed in low-ovo solution because the cells survive trituration better in a lower protein solution. High-ovo is subsequently used to fully quench any residual papain enzymatic activity.

28. Dissociate the brain tissue as follows (see online Movie 2 at cshprotocols.cshlp.org).
    i. Gently pipette the tissue six to eight times through a 5-mL pipette.
    ii. Let the tissue chunks settle for 1–2 min.
    iii. Remove 1–1.5 mL of the resulting cell suspension, trying to avoid large tissue chunks. Place the collected cell suspension in a new, sterile 15-mL conical tube.
    iv. Add 1 mL of low-ovo solution with DNase I, and repeat Steps 28.i–28.iii.
    v. Add 1 mL of low-ovo solution with DNase I, and then follow Steps 28.i–28.iv, now triturating tissue through a 1-mL pipette tip. Repeat until all of the tissue is dissociated and the low-ovo solution prepared in Step 22 is used up.
29. (*Optional*) To determine cell viability and the effectiveness of the dissociation at this stage, remove a small aliquot (50–100 µL) of cell suspension for counting. Count as described in Step 61.

    See Troubleshooting.

30. Pellet the cells at ~220*g* for 15 min in a tabletop centrifuge at room temperature.
31. Aspirate the supernatant, being careful not to disturb the cell pellet.
32. Resuspend the cell pellet in 6 mL of high-ovo solution prepared in Step 11.
33. Immediately pellet the cells at ~220*g* for 15 min in a tabletop centrifuge at room temperature.

    Leaving cells for an extended period of time in high-ovo solution will reduce cell viability.

34. Aspirate the supernatant, being careful not to disturb the cell pellet.
35. Resuspend the cell pellet in 6 mL of panning buffer prepared in Step 11.
36. Prewet a sterile Nitex mesh filter by filtering 2 mL of panning buffer into a sterile 50-mL tube. Filter the cells into the tube 1 mL at a time, then rinse the filter with the remaining panning buffer.

Panning

Do not allow plates to dry out at any stage.

37. Immediately before panning dish use, rinse the panning dish three times with D-PBS.
38. Add the cell suspension from Step 36 to the first rinsed BSL1 dish prepared in Step 6.
39. Incubate the plate for 15 min at room temperature, agitating the plate at 5-min intervals to ensure that all cells have an opportunity to adhere to the plate surface.
40. Gently shake the plate to loosen nonadherent cells, and transfer the cell suspension to the second rinsed BSL1 dish prepared in Step 6.
41. Repeat Step 39.
42. Gently shake the plate to loosen nonadherent cells, and transfer the cell suspension to the first rinsed positive selection dish (e.g., anti-PDGFRa or anti-GalC) prepared in Step 7.
43. Incubate the plate for 45 min at room temperature, agitating the plate every 15 min to ensure that all cells have an opportunity to adhere to the plate surface.
44. Shake the plate to loosen nonadherent cells.

    If subsequent panning steps are to be performed, cell suspension should be kept and transferred to the next panning dish. If only one positive selection is to be used, the cell suspension can be discarded.

Cite this protocol as *Cold Spring Harb Protoc*; doi:10.1101/pdb.prot073973

45. Rinse the plate six times with D-PBS, shaking the plate at each rinse to loosen nonadherent cells.

    OPCs often bind relatively weakly to the anti-PDGFRa plate, so wash gently, pouring the D-PBS onto the same place on the plate each time.

46. For each positive selection plate, repeat Steps 43–45.

    The subsequent trypsinization or lysing steps can be carried out during the 45 min incubation for the next plate.

47. Proceed to Step 48 if cells are to be cultured. Alternatively, to lyse cells for biochemistry, remove all liquid after the final wash. Add RNA or protein lysis buffer directly to the panning dish on ice. Scrape the lysates with a rubber policeman and transfer to a 1.5-mm microcentrifuge tube for further processing.

Trypsinization

See online Movie 3 at cshprotocols.cshlp.org for an illustration of Steps 54 and 55.

48. Before trypsinization, confirm visually under the microscope that nearly all nonadherent cells have been removed by rinsing. If needed, perform additional rinsing steps.

    There will always be a small number (<0.5%) of dislodged floating cells at the end of the rinsing steps, especially when using anti-PDGFRa.

    See Troubleshooting.

49. Once the cells are in their final D-PBS rinse, transfer 4 mL of the equilibrated EBSS prepared in Step 9 to a sterile tube and add 400 µL of trypsin stock solution.

50. Remove the D-PBS from the plate and rinse the plate with the remaining 6 mL of equilibrated EBSS.

51. Pour off the EBSS and add the 4 mL of trypsin-EBSS solution prepared in Step 49.

52. Incubate the plate for 6–8 min in a 37°C incubator.

    Optimal trypsinization time should be determined with each batch of stock trypsin solution generated, and should correspond to the time at which a large percentage of cells on the plate can be easily released by gentle pipetting. Trypsinization times that are either too short or too long can result in reduced cell viability.

53. Add 2 mL of the filtered 30% FCS solution prepared in Step 11 to the plate to stop trypsin digestion.

54. Dislodge the cells from the plate surface by squirting the FCS solution around the plate using a 1-mL pipette tip: Squirt the solution once around the entire circumference toward the center of the dish, once along the edge of the entire dish, and then once concentrating on center. Avoid scraping the tip of the pipette along the surface of the plate; instead squirt just above the surface. Avoid generating excess bubbles in the solution during squirting.

55. Remove the cell suspension from the plate and place it in a sterile 15-mL conical tube.

56. Place 5 mL of fresh 30% FCS in the plate and visualize the plate under the microscope to determine if there are particular regions of the plate that still contain adherent cells.

    Adherent cells are often found along the edges or in the exact center of the plate.

57. Repeat Steps 54 and 55 to dislodge and collect the remaining cells.

58. Rinse the plate with the remaining FCS and add the solution to the collection tube to collect the last remaining cells.

59. Remove a 50- to 100-µL aliquot of the cell suspension for counting.

60. Pellet the cells at ~220*g* for 15 min in a tabletop centrifuge at room temperature.

61. During Step 60, determine yield by adding an equal volume of trypan blue to the cell suspension aliquot from Step 59 and count the cells on a hemocytometer slide.

    Expected yields are as follows:

    One P6-8 mouse brain, 0.5 × 10^6 PDGFRa$^+$ OPCs

One P12 mouse brain, 1×10^6 $GalC^+$ OLs

See Troubleshooting.

62. After centrifugation, aspirate the supernatant and resuspend the cell pellet in a small volume of DMEM-SATO base growth medium.

Plating

OPCs are typically plated at a density of 10,000–20,000 /coverslip in 24-well plate wells, 500,000 per 80-cm^2 flask or 1×10^6 cells per 175-cm^2 flask for proliferation. Mature cells are usually plated at a density of 40,000–50,000/ coverslip in 24-well plate wells.

63. Preplate the cells as follows.

    Coverslips

    i. Adjust the volume of the cell suspension in DMEM-SATO base growth medium to the number of desired cells per well/50 µL medium.
    ii. Place a 50-µL spot of cell suspension at the center of the individual coverslips prepared in Step 3.
    iii. Incubate for 20–45 min at 37°C to let the cells adhere to the coverslips.
    iv. Add the desired medium at 500 µL/well (see Step 64).

    Tissue Culture Plates

    i. Adjust the volume of the cell suspension in DMEM-SATO base growth medium to the number of desired cells per plate/300 µL medium.
    ii. Pipette 300 µL of cell suspension into a 10-cm PDL-coated tissue culture plate prepared in Step 2 and carefully spread the liquid with a sterile glass spreader, trying to avoid scraping the glass spreader on the plastic bottom of the plate.
    iii. Incubate for 7 min at 37°C to let the cells adhere to plate.
    iv. Add the desired medium at 10 mL/plate (see Step 64).
    v. Alternatively, add the number of cells to be plated to 10 mL of the desired medium and add the solution directly to the PDL-coated 10-cm tissue culture plate.

        This method results in slightly lower initial cell viability.

64. Prepare OPC culture medium according to the cells used and the desired result.

    The addition of PDGF, NT-3 and/or T3 may be added to the culture medium to promote proliferation and/or differentiation (Fig. 2).

65. Incubate the OPC cultures at 37°C, 10% CO_2; replace 50% of the medium with fresh medium every 2–3 d.

    To generate highly dense cultures of OPCs, we recommend using 20 ng/mL of PDGF (2× normal concentration) as the OPC cultures begin to get dense (>25% confluency) to avoid unwanted differentiation. In addition, supplemental PDGF can be added directly to the cultures between feedings.

    If differentiating OPCs into mature oligodendrocytes via removal of PDGF and addition of T3, a more rapid and synchronized differentiation can be obtained by passaging the cells into a new flask, rather than simply replacing the medium.

    See Troubleshooting.

TROUBLESHOOTING

Problem (Step 48): Microglial contamination is present.

Solution: Microglial contamination is caused by insufficient depletion of microglia by the BSL1 dishes. Microglia are usually easily identifiable on the positive selection dishes because of their "poached egg" appearance, in contrast to the rounded OPCs or oligodendrocytes. To avoid

Cite this protocol as *Cold Spring Harb Protoc*; doi:10.1101/pdb.prot073973

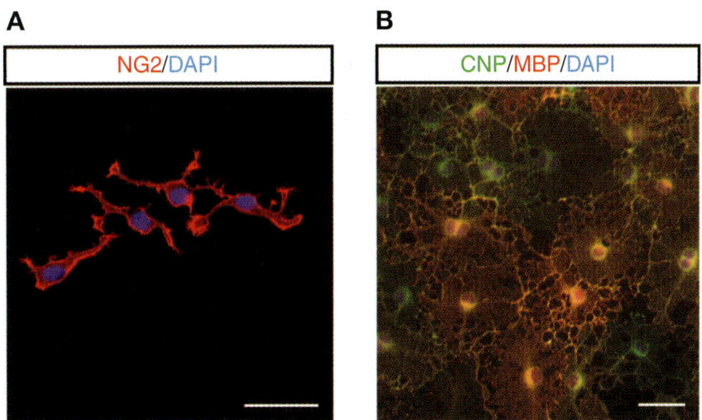

FIGURE 2. Mouse oligodendrocyte lineage cells in culture. (*A*) OPCs purified via immunopanning for PDGFRa and cultured in the presence of PDGF. (*B*) Differentiating oligodendrocytes 48 h following removal of PDGF and addition of T3. Cells have dramatically changed their morphology, and they all express CNP and are in the process of up-regulating MBP. NG2, NG2 chondroitin sulfate proteoglycan; CNP, 2′,3′-cyclic-nucleotide 3′-phosphodiesterase; MBP, myelin basic protein. Scale bars, 10 µm.

contamination, either decrease the number of brains used or increase the number or size of the BSL1 dishes. Petri dishes of 15 cm instead of 10 cm can be used to increase the surface area (while increasing the volume of panning buffer to 15–20 mL), or additional BSL1 dishes can be added. If more dishes are added, reduce the total panning time by splitting the cell suspension into two and passing it over two BSL1 dishes in parallel before combining for the final positive selection dishes.

Problem (Steps 29 and 61): The final yield of cells is low.

Solution: Check the cell count of the total brain supernatant collected in Step 29. For whole brain, the yield at this step should be 20×10^6 total cells. If the yield is much lower, the dissociation steps (Steps 20–28) may have been the problem: Either the tissue was not sufficiently dissociated (if there are very few cells), or the trituration was too rough (if there is a very high percentage of trypan blue–positive dead cells). In addition, if the papain enzyme is bad, dissociation will be unsuccessful.

Problem (Steps 48 and 61): The final yield of cells is low and the cells were not very dense on the final planning plate at Step 48.

Solution: If the dissociation count in Step 29 is satisfactory, the next most likely source of the problem is the panning plates. If the cells were not very dense on the final panning plate at Step 48, there are several potential reasons.

- The cells were unhealthy and did not stick to the plate, possibly because of subtle problems in the trituration steps.
- The cells were rinsed too robustly. (This is especially relevant for anti-PDGFRa plates, but less so for anti-GalC or anti-MOG plates.)
- The panning buffer was made with D-PBS lacking Ca^{2+}/Mg^{2+}, or insulin was not added to the panning buffer. Either of these errors will lead to the cells not adhering strongly to the panning plates.
- The protocol was initiated using too much tissue and, thus, there was too much tissue in the panning plates. Excessive cells will block the surface of the plate and prevent the target cells from contacting and adhering to the plate surface.

Problem (Step 61): The final yield of cells is low, but the pretrypsinization density of cells on the final panning plate was high.

Solution: If the pretrypsinization density of cells was high at Step 48, then the trypsinization and final rinsing step might be a problem.

- If the trypsin solution is too weak, or if the trypsinization time is too short, it is very difficult to dislodge the cells from the final panning plate. In addition, cells that are dislodged will be unhealthy, leading to a low final yield. If after 10 min the cells are still stuck very strongly to the plate, try adding more trypsin to the plate or making a fresh batch of trypsin.
- Alternatively, greatly overtrypsinizing the cells will also be detrimental to cell health, so if the cells come off very easily with one squirt, try reducing the trypsinization time slightly.

Problem (Step 65): Cells exhibit poor viability the day after plating.
Solution: This problem is often associated with the PDL substrate. Try coating with PDL overnight at 4°C and ensure that the dishes dry well between the washes and the addition of cells. Alternatively, the cells may have been over- or undertrypsinized. If the cells were difficult to remove from the panning dishes, try increasing the trypsin concentration or time. Conversely, if the cells came off the dish extremely easily, try decreasing the trypsin concentration or time.

RECIPES

CNTF Stock (10 μg/mL)

To prepare, dilute ciliary neurotrophic factor (CNTF; Peprotech 450-02) to 10 μg/mL with sterile 0.2% BSA in Dulbecco's phosphate-buffered saline (D-PBS; HyClone SH30264.01). Make 20-μL aliquots, flash freeze in liquid nitrogen, and store –80°C.

DMEM-SATO Base Growth Medium

1. Combine the following:

Reagent	Amount (for 20 mL)	Final concentration
Dulbecco's Modified Eagle's Medium (Invitrogen 11960-044)	19.5 mL	1×
SATO supplement (100×) <R>	200 μL	1×
Glutamine (200 mM; Invitrogen 25030-081)	200 μL	2 mM
Penicillin-streptomycin (Gibco/Life Technologies 15140-122)	200 μL	100 U/mL (penicillin)100 μg/mL (streptomycin)
Sodium pyruvate (100 mM; Invitrogen 11360-070)	200 μL	1 mM
Insulin stock (0.5 mg/mL) <R>	200 μL	5 μg/mL
N-Acetyl-L-cysteine (Sigma-Aldrich A8199) stock (5 mg/mL, prepared in DMEM)	20 μL	5 μg/mL
Trace Elements B (1000×; Cellgro 99-175-CI)	20 μL	1×
d-Biotin (Sigma-Aldrich B4639) stock (50 μg/mL)	4 μL	10 ng/mL

2. To improve viability of mouse OPCs and oligodendrocytes, include 400 μL of 50× B-27 (Invitrogen 17504-044).

   SM1 supplement (Stemcell Technologies 05711) or NS21 supplement (Chen et al. 2008; R&D Systems AR008) may be used in place of B-27 to promote survival. Note that these commercially available supplements all contain small but biologically relevant amounts of T3. We have not found that this prevents expansion of cortical mouse OPCs for up to at least 2 wk in culture, however; if you wish to study OPCs in the absence of T3-mediated signaling, prepare NS21 lacking T3 as per Chen et al. (2008) or use rat OPCs (which do not require the supplements for survival).

3. Filter through a 0.22-µm filter to sterilize. Store at 4°C for up to 1 wk.

DNase I

On ice, dissolve 12,500 U of DNase I (Worthington Biochemical LS002007) per 1 mL of chilled Earle's Balanced Salt Solution (EBSS; Invitrogen 14155-063). Filter-sterilize on ice. Aliquot (e.g., 200 µL/tube) and freeze overnight at −20°C. Store aliquots at −20°C to −30°C.

EBSS Stock (10×)

Reagent	Quantity (for 250 mL)	Final concentration (10×)
NaCl	17 g	1.16 M
KCl	1 g	54 mM
$NaH_2PO_4 \cdot H_2O$	0.35 g	10 mM
Glucose	2.5 g	1%
Phenol red (0.5%)	2.5 mL	0.005%

Bring to 250 mL with ddH_2O and filter to sterilize.

Ethanol-Washed Glass Coverslips

Extensively wash 12-mm glass coverslips (Carolina Biological Supply 633029) in 70% ethanol. Perform the washes on a platform shaker in a beaker, with enough motion to lightly agitate the coverslips but not break too many. Wash the coverslips for about 1 mo, exchanging the ethanol approximately every day. (It is fine to skip some exchanges.) Store the washed coverslips in 70% ethanol until use.

Forskolin Stock (4.2 mg/mL)

To prepare, add 1 mL of sterile DMSO to a 50-mg bottle of forskolin (Sigma-Aldrich F6886) and pipette up and down until the powder is fully resuspended. Transfer to a 15-mL conical tube and add an additional 11 mL of DMSO to achieve a final concentration of 4.2 mg/mL. Store in 20- and 80-µL aliquots at −20°C.

High-Ovomucoid Stock (6×)

1. Dissolve the following in 160–180 mL of Dulbecco's phosphate-buffered saline (D-PBS; Invitrogen 14287-080).

Reagent	Amount (for 200 mL)	Final concentration
BSA (Sigma_Aldrich A-8806)	6 g	30 mg/mL
Trypsin inhibitor (Worthington LS003086)	6 g	30 mg/mL

2. Adjust the pH to 7.4 with 10 N NaOH. Bring the volume to 200 mL with D-PBS, and then filter-sterilize.
3. Make 1-mL aliquots and store them at −20°C.

Insulin Stock (0.5 mg/mL)

To 20 mL of sterile water, add 10 mg of insulin (Sigma-Aldrich I6634) and 100 µL of 1.0 N HCl. Mix well. Filter through a 0.22-µm filter. Store at 4°C for 4–6 wk.

Low-Ovomucoid Stock Solution (10×)

To prepare, add 3 g of BSA (Sigma-Aldrich A8806) to 150 mL D-PBS. Mix well. Add 3 g of trypsin inhibitor (Worthington LS003086) and mix to dissolve. Add ~1 mL of 1 N NaOH to adjust the pH to 7.4. Bring the volume to 200 mL with D-PBS. Filter-sterilize through a 0.22-µm filter. Make 1.0-mL aliquots and store at −20°C.

NT-3 Stock (1 µg/mL)

1. Prepare a master stock of neurotrophin-3 (NT-3; Peprotech 450-03) at 1–0.1 mg/mL according to manufacturer's instructions. (Instructions may vary from lot to lot; generally, NT-3 is dissolved in buffer [e.g., Dulbecco's phosphate-buffered saline] plus 0.2% BSA.) Store at −80°C.
2. Prepare 0.2% BSA (Sigma-Aldrich A4161) in Dulbecco's phosphate-buffered saline (D-PBS; Invitrogen 14287-080). Filter-sterilize and chill.
3. Dilute an aliquot of NT-3 master stock to 1 µg/mL in sterile, chilled 0.2% BSA prepared in Step 2.
4. Aliquot the NT-3 working stock from Step 3 (e.g., 20 µL/tube) and snap-freeze in liquid nitrogen.
5. Store aliquots at −80°C.

OPC Culture Medium

1. To prepare full OPC culture medium, combine the following:
   20 mL DMEM-SATO base growth medium <R>
   20 µL Forskolin stock (4.2 mg/mL) <R>
   20 µL CNTF stock (10 µg/mL) <R>
2. Add growth factors as appropriate.
   - To promote OPC proliferation, omit PDGF and add:
     20 µL PDGF stock (10 µg/mL) <R>
     20 µL NT-3 stock (1 µg/mL) <R>
   - To promote OL differentiation, omit PDGF and add:
     200 µL T3 stock (4 µg/mL) <R>
3. Store at 4°C for up to 3 d.

Papain Buffer

Reagent	Amount (for 250 mL)	Final concentration
EBSS stock (10×) <R>	25 mL	1×
$MgSO_4$ (100 mM)	2.5 mL	1 mM
Glucose (30%)	3 mL	0.46 %
EGTA (0.5 M)	1 mL	2 mM
$NaHCO_3$ (1 M)	6.5 mL	26 mM

Bring volume up to 250 mL with ddH_2O and filter to sterilize.

Cite this protocol as *Cold Spring Harb Protoc*; doi:10.1101/pdb.prot073973

PDGF Stock (10 µg/mL)

1. Prepare a platelet-derived growth factor (PDGF) master stock by diluting PDGF (Peprotech 100-13A) at 1–0.1 mg/mL according to manufacturer's instructions. (Instructions may vary from lot to lot; generally, PDGF is dissolved in buffer [e.g., Dulbecco's phosphate-buffered saline] plus 0.2% BSA.) Store at −80°C.
2. Prepare 0.2% BSA (Sigma-Aldrich A4161) in Dulbecco's phosphate-buffered saline (D-PBS; Invitrogen 14287-080). Filter-sterilize and chill.
3. Dilute an aliquot of PDGF master stock to 10 µg/mL in sterile, chilled 0.2% BSA prepared in Step 2.
4. Aliquot the PDGF working stock from Step 3 (e.g., 20 µL/tube) and snap-freeze in liquid nitrogen.
5. Store aliquots at −80°C.

Poly-D-Lysine (PDL) Stock (1 mg/mL)

1. Resuspend poly-D-lysine (PDL; Sigma-Aldrich P6407; molecular weight 70–150 kDa) at 1 mg/mL in borate buffer by combining 50 mg of PDL and 50 mL of 0.15 M boric acid (pH 8.4).
2. Filter to sterilize, and then aliquot (e.g., 100 µL/tube). Store aliquots at −20°C.

SATO Supplement (100×)

1. Prepare the following stock solutions (these should be made fresh; do not reuse):
   - Combine 5 mg of progesterone (Sigma-Aldrich P8783) and 200 µL of ethanol to make a progesterone stock solution.
   - Combine 4 mg of sodium selenite (Sigma-Aldrich S5261), 10 µL of 1 N NaOH, and 10 mL of Dulbecco's modified Eagle's medium (DMEM; Gibco/Life Technologies 11960-044) to make a sodium selenite stock solution.
2. Combine the following:

Reagent	Quantity (for 200 mL)	Final concentration (100×)
BSA (Sigma-Aldrich A4161)	2 g	10 mg/mL
Transferrin (Sigma-Aldrich T1147)	2 g	10 mg/mL
Putrescine (Sigma-Aldrich P5780)	320 mg	1.6 mg/mL
Progesterone stock solution	50 µL	6 µg/mL
Sodium selenite stock solution	2 mL	4 µg/mL

3. Bring to a total volume of 200 mL in DMEM, and then filter-sterilize. Aliquot and store at –20°C.

T3 Stock (4 µg/mL)

1. Dissolve 4 µg of thyroid hormone (T3; Sigma-Aldrich T6397) in 500 µL of 1 N NaOH to prepare a solution of 0.8 µg/100 µL.
2. Add 75 µL of the T3 solution from Step 1 to 150 mL of Dulbecco's phosphate-buffered saline (D-PBS; Invitrogen 14287-080).
3. Filter solution through a filter-sterilization unit, discarding the first 10 mL.
4. Aliquot (e.g., 200 µL/tube) and then store at −20°C.

REFERENCE

Chen Y, Stevens B, Chang J, Milbrandt J, Barres BA, Hell JW. 2008. NS21: Re-defined and modified supplement B27 for neuronal cultures. *J Neurosci Methods* **171:** 239–247.

Cite this protocol as *Cold Spring Harb Protoc*; doi:10.1101/pdb.prot073973

CHAPTER 9

Myelinating Cocultures of Purified Oligodendrocyte Lineage Cells and Retinal Ganglion Cells

Trent A. Watkins[1,2,3] and Anja R. Scholze[1]

[1]*Department of Neurobiology, Stanford University School of Medicine, Stanford, California 94305-5125*

In this chapter, we introduce methods for generating rapidly myelinating cocultures with reaggregates of purified retinal ganglion cells and optic nerve oligodendrocyte precursor cells. This coculture system facilitates the study of complex central nervous system neuronal–glial interactions and myelination. It enables control of the extracellular environment and allows the use of transfected, virally infected, mutant, or knockout neurons and/or glial cell types. It is therefore possible to assess the role of various signaling pathways and genes in myelination and node of Ranvier formation.

INTRODUCTION

The axonal–glial interactions that result in myelination in the central nervous system (CNS) are poorly understood, as are the processes underlying the wrapping of axons to form compact myelin. Although culture of isolated oligodendrocyte (OL) lineage cells has yielded a number of important insights into the development and differentiation of this lineage, the approach is limited in its utility for studies of neuronal–glial interactions. Methods for adapting the culture of purified retinal ganglion cells (RGCs) for myelinating coculture with purified OL lineage cells are introduced in this chapter. This system helps to bridge a gap between the simplified in vitro study of purified cells and the complex cellular interactions that define CNS myelination in vivo. The approach and its applications have been described previously (Watkins et al. 2008) and are adapted here.

ESTABLISHING THE COCULTURE

Early efforts at coculturing dissociated RGCs with oligodendrocyte precursor cells (OPCs) did not result in myelination, even after prolonged periods (Meyer-Franke et al. 1999). Several modifications have been developed to enable the formation of compact myelin in these cocultures (Watkins et al. 2008). First, the architecture of the coculture has been adjusted to use clusters, or reaggregates, of RGCs. Second, modification of the culture medium has enabled more robust wrapping of axons with compact myelin. Finally, various chemical and cellular additives that can influence distinct stages of myelination have been identified. Together, these modifications enable a rapid and versatile system for studying CNS myelination and are detailed in Protocol 1: Myelinating Cocultures of Rat Retinal Ganglion Cell Reaggregates and Optic Nerve Oligodendrocyte Precursor Cells (Watkins and Scholze).

[2]Present address: Department of Neuroscience, Genentech Inc., South San Francisco, California 94080.

[3]Correspondence: watkins.trent@gene.com

Cite this introduction as *Cold Spring Harb Protoc*; doi:10.1101/pdb.top070839

The adapted RGC reaggregate architecture allows for the generation of dense beds of axons that serve as a substrate for the seeding of purified OL lineage cells. Reaggregates of purified postnatal rodent RGCs are achieved by plating freshly isolated neurons at high density on a substrate to which they attach poorly, thereby causing them to self-adhere instead. Upon plating, each RGC reaggregate extends a multitude of axons radially to generate a dense array of aligned axons. This neuronal architecture sharply contrasts with the diffuse network of unaligned neurites generated in dissociated RGC cultures and promotes more robust myelination by OPCs upon coculture (Fig. 1).

The choice of medium can influence the interactions of glia and neurons in coculture. Slight modification of standard RGC growth medium is sufficient to promote the survival and growth of both RGCs and OPCs, including mouse glia, as well as to support OL ensheathment of axons. This medium, however, does not strongly support wrapping of axons to form compact myelin, a limitation that prompted the formulation of a new coculture medium, termed myelination medium (MyM). Compared to RGC growth medium, MyM contains reduced concentrations of many supplements, including B-27, and includes the addition of vitamin B12, hydrocortisone, and ceruloplasmin. Cocultures grown in MyM show more robust total myelination, as well as greater myelin wrapping and compaction.

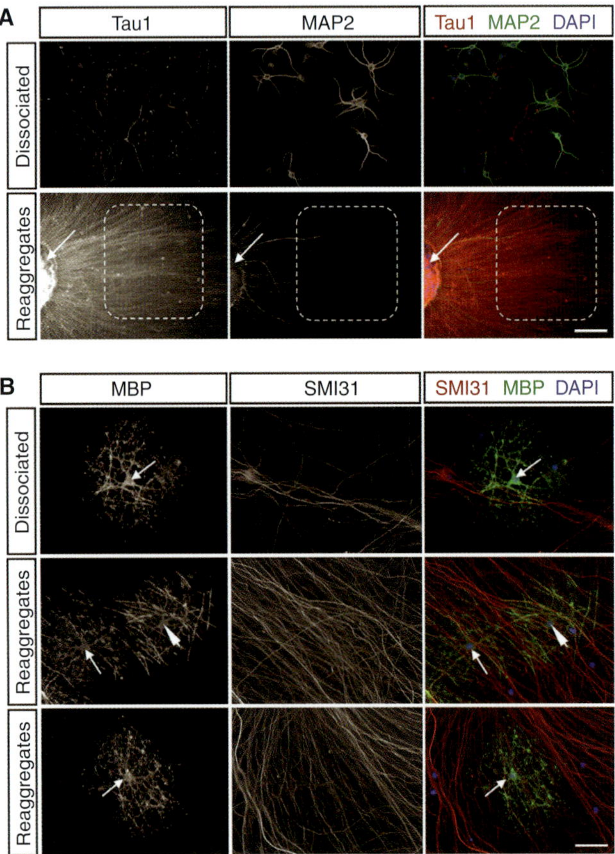

FIGURE 1. Reaggregates of RGCs extend dense bed of axons in culture that serve as a substrate for myelination by OLs. (*A*) Immunostaining of 10-d cultures of dissociated RGCs or RGC reaggregates for markers of axons (Tau1) and dendrites (MAP2). Cultures of dissociated RGCs have neuronal cell bodies and dendrites distributed throughout the assortment of neurites, whereas an RGC reaggregate (arrow) extends dendrites only a limited distance, leaving dense regions of axons running roughly in parallel (dotted box). Scale bar, 50 µm. (*B*) Immunostaining of 6-d cocultures of optic nerve OPCs and RGC reaggregates for MBP and neurofilament (SMI31). The increased axon density of RGC reaggregates is associated with the appearance of myelinating OLs (arrowhead), but the majority of MBP-expressing OLs in axon-dense regions still failed to form clear myelin segments around axons (arrows). (Scale bar, 50 µm.) (Reprinted, with permission, from Watkins et al. 2008.)

 Cite this introduction as *Cold Spring Harb Protoc*; doi:10.1101/pdb.top070839

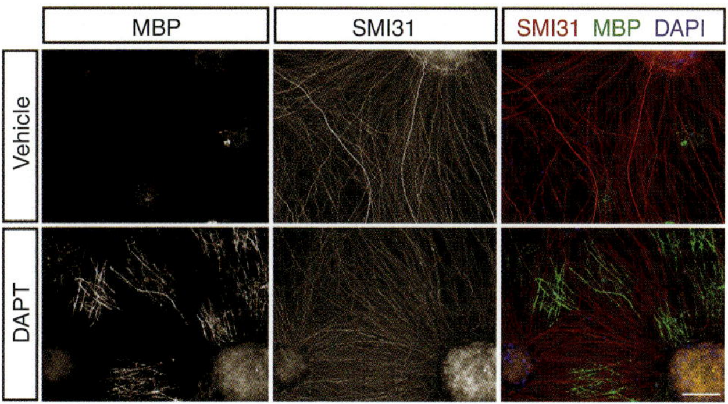

FIGURE 2. Treatment of cocultures with a γ-secretase inhibitor disinhibits OL differentiation and myelination. Immunostaining of 4-d DAPT-treated cocultures reveals substantial OL differentiation and myelination. (Scale bar, 100 μm.) (Reprinted, with permission, from Watkins et al. 2008.)

The degree of ensheathment and wrapping of RGC axons by differentiating OLs can be influenced by other adaptations to the coculture system (Watkins et al. 2008). Addition of γ-secretase inhibitors (Dovey et al. 2001) to cocultures improves both differentiation of OPCs (by blocking Notch1 signaling) and ensheathment of axons (by an unknown Notch1-independent mechanism) (Fig. 2) (Wang et al. 1998; Genoud et al. 2002; Watkins et al. 2008). The addition of immunopanned rat optic nerve astrocytes (Meyer-Franke et al. 1999; Mi and Barres 1999) stimulates myelin wrapping, resulting in greater myelin thickness. (Interestingly, we have observed that, in contrast, immature cortical astrocytes purified by the McCarthy–de Vellis method [McCarthy and de Vellis 1980] are, for unknown reasons, incompatible with the survival and maintenance of RGC reaggregates and are therefore not an appropriate substitute for promoting myelination.) Robust generation of compact myelin in cocultures can therefore be achieved by addition of both γ-secretase inhibitors and optic nerve astrocytes.

MATURATION OF THE COCULTURE

The development of myelinating OLs from OPCs can be observed within the first week of seeding OPCs onto RGC axons. In a typical coculture, RGC reaggregates are allowed to extend axons for 1 wk, after which purified rat optic nerve astrocytes may be added along with a partial medium change to MyM. A day or two later, rat optic nerve OPCs are seeded in MyM, including 1 μM DAPT, if desired. After only 3 d, there are substantial numbers of OLs expressing myelin basic protein. Compact myelin can be seen by the sixth day of coculture. This rapid terminal differentiation allows for the effective transient transfection of OPCs with either siRNAs or plasmids before plating (Watkins et al. 2008).

RGC–OPC cocultures are useful for studying not only myelination but also the control of OL differentiation by CNS axons. Unlike the nearly uniform differentiation of isolated OPCs to OLs in the absence of added mitogens, OPCs seeded onto RGC axons have a variety of fates, with axons both maintaining undifferentiated OPCs and inducing differentiation into type 2 astrocytes (Fig. 3). The ratios and mechanisms vary depending on whether the OPCs are isolated from mouse or rat and from brain or optic nerve (Watkins et al. 2008). In the case of OPCs purified from rat optic nerve, approximately equal portions will (1) differentiate into OLs, (2) differentiate into type 2 astrocytes, or (3) remain OPCs but develop a multipolar morphology. In the presence of γ-secretase inhibitors, however, the large majority of rat optic nerve OPCs differentiates into myelinating OLs. Rat brain OPCs display similar ratios when seeded onto RGC axons, but inhibition of γ-secretase only modestly affects differentiation of these cells. With mouse brain OPCs, a greater proportion become astrocytes, with less than one-quarter typically differentiating to OLs. Beyond Notch signaling, the full range of mechanisms that RGC axons use to influence OPC cell fate has not been elucidated.

A primary advantage of this myelinating coculture system is its versatility. Unlike mixed brain myelinating cultures or slice cultures (Lubetzki et al. 1993; Notterpek et al. 1993; Thomson et al.

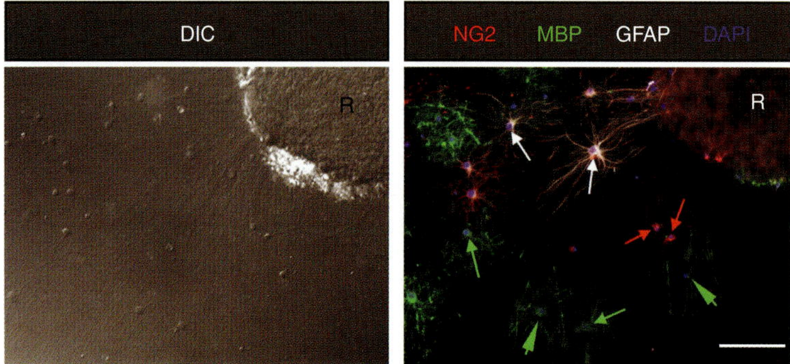

FIGURE 3. Cell fate and morphology in a coculture 6 d after seeding optic nerve OPCs (*left*, differential-interference contrast [DIC] microscopy; *right*, indirect immunofluorescence). Examples of OPCs (NG2, red), nonmyelinating OLs (MBP, green), and astrocytes (GFAP, white) are indicated by arrows of the corresponding colors. Myelinating OLs (green arrowheads) are identified by their extension of multiple distinctive smooth tubes. R, RGC reaggregate. Scale bar, 100 µm. (Reprinted, with permission, from Watkins et al. 2008.)

2006), the choice of starting cells can be made according to the demands of particular experimental questions. For robust differentiation and ensheathment of axons, one may begin with rat optic nerve OPCs cultured in the presence of DAPT. For greater wrapping, optic nerve astrocytes may be added. When ease and yield are priorities, brain OPCs can be obtained in much larger numbers and their propensity to differentiate into type 2 astrocytes on RGC axons can provide improved wrapping without the need for isolating optic nerve astrocytes. One may also evaluate the capacity of various developmental stages of the OL lineage, such as adult OPCs (Shi et al. 1998) or mature OLs (Cahoy et al. 2008), to myelinate under various conditions. Moreover, the ability to use RGCs or OPCs from transgenic mice enables a variety of mechanistic studies. Mixing and matching purified CNS cells therefore provides a valuable tool for a variety of studies of axonal–glial interactions.

REFERENCES

Cahoy JD, Emery B, Kaushal A, Foo LC, Zamanian JL, Christopherson KS, Xing Y, Lubischer JL, Krieg PA, Krupenko SA, et al. 2008. A transcriptome database for astrocytes, neurons, and oligodendrocytes: A new resource for understanding brain development and function. *J Neurosci* **28:** 264–278.

Dovey HF, John V, Anderson JP, Chen LZ, de Saint Andrieu P, Fang LY, Freedman SB, Folmer B, Goldbach E, Holsztynska EJ, et al. 2001. Functional γ-secretase inhibitors reduce β-amyloid peptide levels in brain. *J Neurochem* **76:** 173–181.

Genoud S, Lappe-Siefke C, Goebbels S, Radtke F, Aguet M, Scherer SS, Suter U, Nave KA, Mantei N. 2002. Notch1 control of oligodendrocyte differentiation in the spinal cord. *J Cell Biol* **158:** 709–718.

Lubetzki C, Demerens C, Anglade P, Villarroya H, Frankfurter A, Lee VM, Zalc B. 1993. Even in culture, oligodendrocytes myelinate solely axons. *Proc Natl Acad Sci* **90:** 6820–6824.

McCarthy KD, de Vellis J. 1980. Preparation of separate astroglial and oligodendroglial cell cultures from rat cerebral tissue. *J Cell Biol* **85:** 890–902.

Meyer-Franke A, Shen S, Barres BA. 1999. Astrocytes induce oligodendrocyte processes to align with and adhere to axons. *Mol Cell Neurosci* **14:** 385–397.

Mi H, Barres BA. 1999. Purification and characterization of astrocyte precursor cells in the developing rat optic nerve. *J Neurosci* **19:** 1049–1061.

Notterpek LM, Bullock PN, Malek-Hedayat S, Fisher R, Rome LH. 1993. Myelination in cerebellar slice cultures: Development of a system amenable to biochemical analysis. *J Neurosci Res* **36:** 621–634.

Shi J, Marinovich A, Barres BA. 1998. Purification and characterization of adult oligodendrocyte precursor cells from the rat optic nerve. *J Neurosci* **18:** 4627–4636.

Thomson CE, Hunter AM, Griffiths IR, Edgar JM, McCulloch MC. 2006. Murine spinal cord explants: A model for evaluating axonal growth and myelination in vitro. *J Neurosci Res* **84:** 1703–1715.

Wang S, Sdrulla AD, diSibio G, Bush G, Nofziger D, Hicks C, Weinmaster G, Barres BA. 1998. Notch receptor activation inhibits oligodendrocyte differentiation. *Neuron* **21:** 63–75.

Watkins TA, Emery B, Mulinyawe S, Barres BA. 2008. Distinct stages of myelination regulated by γ-secretase and astrocytes in a rapidly myelinating CNS coculture system. *Neuron* **60:** 555–569.

Watkins TA, Scholze AR. 2014. Myelinating cocultures of rat retinal ganglion cell reaggregates and optic nerve oligodendrocyte precursor cells. *Cold Spring Harb Protoc* doi:10.1101/pdb.prot074971.

 Cite this introduction as *Cold Spring Harb Protoc*; doi:10.1101/pdb.top070839

Protocol 1

Myelinating Cocultures of Rat Retinal Ganglion Cell Reaggregates and Optic Nerve Oligodendrocyte Precursor Cells

Trent A. Watkins[1,2,3] and Anja R. Scholze[1]

[1]*Department of Neurobiology, Stanford University School of Medicine, Stanford, California 94305-5125*

This protocol describes the generation of a rapidly myelinating central nervous system coculture for the study of complex neuronal-glial interactions in vitro. Postnatal rat retinal ganglion cells (RGCs) purified by immunopanning are promoted to cluster into reaggregates and then allowed to extend dense beds of radial axons for 10–14 d. Subsequently, rodent oligodendrocyte precursor cells are purified by immunopanning, transfected if desired, and seeded on top of the RGC reaggregates. Under the conditions described here, compact myelin can be observed within 6 d.

MATERIALS

It is essential that you consult the appropriate Material Safety Data Sheets and your institution's Environmental Health and Safety Office for proper handling of equipment and hazardous material used in this protocol.

RECIPES: Please see the end of this protocol for recipes indicated by <R>. Additional recipes can be found online at http://cshprotocols.cshlp.org/site/recipes.

Reagents

BSA stock (4%) <R>

DNase

To prepare a 0.4% stock of DNase in Earle's balanced salt solution (EBSS), add 1 mL of Earle's balanced salt solution (EBSS; Sigma-Aldrich E6267) per 12,500 U of DNase I (Worthington LS002007). Keep on ice. Filter sterilize, and store in 200-µL aliquots at −20°C.

Dulbecco's phosphate-buffered saline (D-PBS, Gibco 14287-080)

Earle's balanced salt solution (EBSS) (Sigma-Aldrich E6267)

Fetal calf serum (FCS) (Gibco 10437-028)

Prepare 50-mL aliquots of FCS. Heat-inactivate aliquots for 30 min at 55°C, and then store at −20°C.

High-ovomucoid (high-ovo) stock solution (6×)<R>

Immunopanning reagents

- A2B5 hybridoma supernatant or ascites
- Galactocerebroside (GC) hybridoma supernatant
- Goat anti-mouse IgG + IgM (H + L) (Jackson ImmunoResearch 115-005-044)

[2]Present address: Department of Neuroscience, Genentech Inc., South San Francisco, California 94080.

[3]Correspondence: Watkins.Trent@gene.com

Cite this protocol as *Cold Spring Harb Protoc*; doi:10.1101/pdb.prot074971

Goat anti-mouse IgM, μ-chain specific (Jackson ImmunoResearch 115-005-020)
Ran-2 hybridoma supernatant

InSolution γ-Secretase Inhibitor I× (DAPT) (Calbiochem/EMD Millipore 565784) (optional; see Step 47)
Insulin stock (0.5 mg/mL) <R>
L-Cysteine hydrochloride monohydrate (Sigma-Aldrich C7880)
Laminin (mouse) (Cultrex; R&D Systems 3400-010-01)

Thaw mouse laminin (1 mg/mL) at 4°C. Make 10-μL aliquots and store at −30°C.

Low-ovomucoid (low-ovo) stock solution <R>
Myelination medium (MyM), freshly prepared <R>
NaOH (1 M)
ND-SATO growth medium (ND-G), freshly prepared <R>
Neurobasal solution (NB) (Gibco 21103-049)
Nucleofector kit for rat glial cells (Lonza) (if performing transfection)
Papain (Worthington LS03126)
Plasmid or siRNA for transfection, as desired
Poly-D-lysine (PDL) (1 mg/mL)
Rat pups (P5, Sprague–Dawley) for purification of RGCs (three litters/30 animals)

P4–P10 pups may be used, but P5 is optimal for its combination of yield, viability, and purity. RGCs are purified using the method described in Chapter 1, Protocol 1: Purification and Culture of Retinal Ganglion Cells from Rodents (Winzeler and Wang). Mouse RGCs can also be used, per the provided modifications.

Rat pups (P7, Sprague–Dawley) for purification of optic nerve OPCs (3 litters/30 animals)

Pups of nearly any postnatal age will work for OPC purification, but P6–P8 produces the best yield. Cortical OPCs can also be used in the myelinating cocultures and the OPC yields are much higher (2–3 million per P7 cortex; see Chapter 8, Protocol 1: Purification of Oligodendrocyte Precursor Cells from Rat Cortices by Immunopanning [Dugas and Emery]). Mouse OPCs can also be used, per the provided modifications.

Tris-HCl (50 mM, pH 9.5; sterile)

Dissolve 12.1 g of Trizma base in 200 mL of dH_2O. Adjust pH to 9.5 with HCl.

Trypsin stock (2.5%)

Dissolve trypsin (Sigma-Aldrich T9935) at 30,000 U/mL in EBSS (Sigma-Aldrich E6267). Filter through a 0.22-μm filter. Make 200-μL aliquots and store at −80°C.

Trypsin-EDTA (Gibco) (if performing transfection)

Equipment

Centrifuge (tabletop, with 15- and 50-mL conical tube adaptors)
Centrifuge (clinical)
Chamber slides (eight well)
Coverslips (glass)

We wash our glass coverslips with at least three washes of 70% ethanol over 7 d on a rotating platform and store in 70% ethanol.

Dissection tools

Forceps
Scissors (fine)
Scissors (sharp)

Incubator (37°C/10% CO_2)
Microscope
Needles (#21 and #23 gauge)
Nylon mesh filter (20 μm) (Nitex; Tetko)
Nucleofector I device (Lonza) (if performing transfection)
Pasteur pipettes
Petri dishes (35, 10, 15 cm)

 Cite this protocol as *Cold Spring Harb Protoc*; doi:10.1101/pdb.prot074971

Syringes (2.5 mL)
Syringe filters (0.22 µm)
Tissue culture dishes (10 cm) (if performing transfection)
Tissue culture plates (24-well)
Tubes (1.5-mL microcentrifuge)
Tubes (15-mL conical)
Water bath (37°C)

METHOD

For a general outline of the coculture procedure, see Figure 1.

Generation of RGC Reaggregates

Day 1

1. Purify RGCs from three litters of P5 Sprague–Dawley rat pups by immunopanning as described in Chapter 1, Protocol 1: Purification and Culture of Retinal Ganglion Cells from Rodents (Winzeler and Wang).
2. Resuspend the purified RGCs in 2 mL of ND-G medium.
3. Divide the RGCs (typically ~2.5 million neurons) into four wells of an eight-well chamber slide.
4. Incubate the cells at 37°C (10% CO_2).

Day 2

5. Gently resuspend the RGCs to promote reaggregate formation by pipetting the medium up and down eight to 12 times per well using a P200 pipette.
6. Return the cells to 37°C (10% CO_2).
7. Coat coverslips with PDL as follows.
   i. Rinse 24 to 32 glass coverslips three times with sterile distilled H_2O.
   ii. Distribute the coverslips individually around a 15-cm Petri dish.
   iii. Aspirate any remaining water droplets.

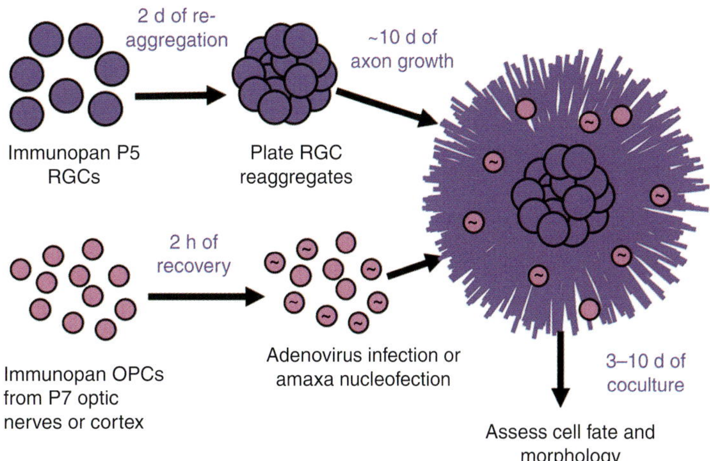

FIGURE 1. Myelinating cocultures of OPCs with RGC reaggregates. RGCs are purified by immunopanning and allowed to reaggregate for 2 d before plating on a substrate permissive for axon growth for 7–14 d. OPCs are then acutely isolated from optic nerves or cortices and transfected if desired. Transfer of these OPCs to the reaggregate cultures initiates the coculture that results in myelination. (Reprinted, with permission, from Watkins et al. 2008.)

iv. Dilute 100 µL of PDL stock in 10 mL of sterile distilled H_2O.

v. Add 100 µL of diluted PDL to the surface of each coverslip. Incubate the coverslips at room temperature for 30 min.

vi. Rinse the coverslips three times with sterile distilled H_2O.

vii. Transfer the coverslips to the wells of 24-well culture plates.

viii. Aspirate the excess H_2O from each well, using the tip of the pipette to position the coverslip in the center of each well.

Day 3

8. Equilibrate 6 mL of ND-G medium to 37°C (10% CO_2) in an incubator.
9. Coat the PDL-coated coverslips with laminin as follows.

    i. Dilute 10 µL of mouse laminin stock into 5 mL of NB.

    ii. Gently add 50 µL of this diluted laminin to each coverslip from Step 7, allowing the surface tension to prevent the drop from expanding off the coverslip to the rest of the well.

    iii. Gently transfer the plate(s) to the incubator.

    iv. Incubate the plate(s) at 37°C (10% CO_2) for 4–6 h.

10. Switch the coverslips to ND-G medium as follows.

    i. Aspirate the laminin solution from each coverslip and immediately replace it with 30 µL of ND-G medium.

    ii. Return the plate(s) to the incubator.

11. Gently resuspend the RGC reaggregates in each well with a P200 pipette and transfer the contents of each well to a 1.5-mL microcentrifuge tube.
12. Wash the RGC reaggregates as follows.

    i. Allow the reaggregates to settle. Remove the supernatant containing debris and unaggregated RGCs.

    ii. Add 500 µL of ND-G medium.

    iii. Allow the reaggregates to settle and remove the supernatant.

    iv. Repeat Steps 12.ii–12.iii five to six times, or until no further debris can be seen in the supernatant under the microscope.

13. Resuspend the RGC reaggregates in 600 µL of ND-G medium.
14. Add 20 µL of medium containing reaggregates to each coverslip from Step 10. Gently distribute the reaggregates around the edge of each coverslip to avoid having the reaggregates cluster in the center.

    The goal is to evenly distribute the reaggregates around the surface of each coverslip. This requires gentle resuspension before plating on each coverslip, as the reaggregates tend to rapidly settle.

15. Gently return the plate(s) to the incubator.

    Excessive motion at this stage can result in spread of the liquid off of the coverslips or swirling of the reaggregates toward the centers of the coverslips, disrupting the even distribution.

Day 4

16. Add 450 µL of ND-G medium to each well of RGCs.
17. Feed the cultures every third day by removing 225 µL of medium and adding 250 µL of fresh equilibrated ND-G medium.

    See note at Step 47.

18. Seed purified OPCs on top of RGC reaggregates (Steps 46–47) after 8–12 d of culture.

Cite this protocol as *Cold Spring Harb Protoc*; doi:10.1101/pdb.prot074971

Isolation of Rat Optic Nerve OPCs by Immunopanning

A schematic diagram of OPC purification is provided in Figure 2.

Day 1

19. Prepare the panning dishes by incubating three 10-cm Petri dishes overnight at 4°C in 10 mL of 50 mM Tris (pH 9.5) plus the following secondary antibodies:

    i. Dish 1: goat anti-mouse IgG (H + L) (10 µg/mL)

    ii. Dish 2: goat anti-mouse IgG (H + L) (10 µg/mL)

    iii. Dish 3: goat anti-mouse IgM (µ-chain specific) (10 µg/mL)

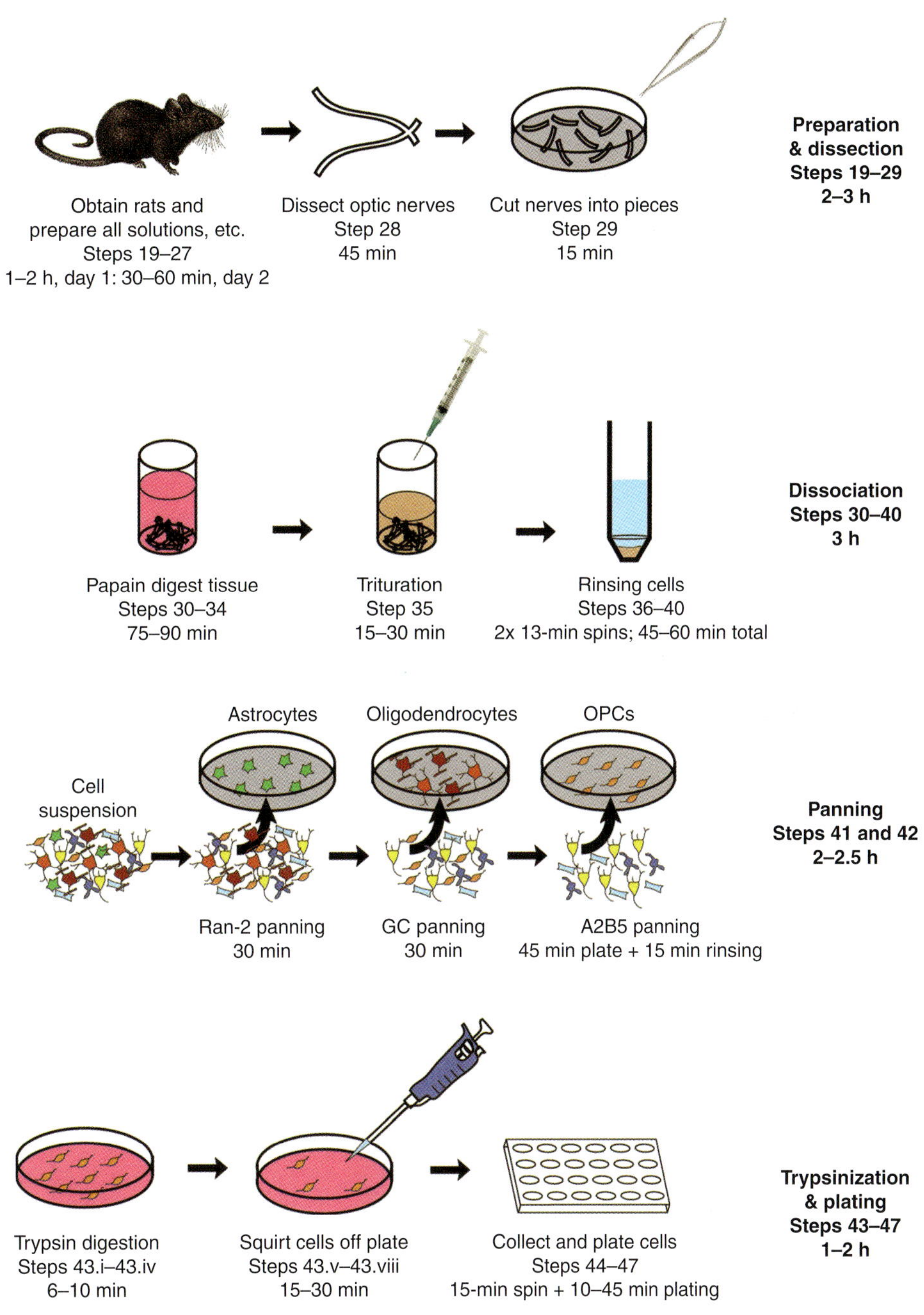

FIGURE 2. Isolation of rat optic nerve OPCs by immunopanning.

Day 2

20. Prepare a D-PBS/0.2% BSA solution by mixing 19 mL of D-PBS with 1 mL of 4% BSA stock.
21. Prepare the panning buffer as follows.
    i. Mix 2 mL of D-PBS/0.2% BSA with 18 mL of D-PBS.
    ii. Add 200 µL of insulin stock.
    iii. Equilibrate to room temperature.
22. Prepare the low-ovo solution as follows.
    i. Mix 1 mL of low-ovo stock solution with 9 mL of D-PBS.
    ii. Add 100 µL of 0.4% DNase stock.
23. Prepare the high-ovo solution by mixing 1.2 mL of high-ovo stock solution with 5 mL of D-PBS.
24. Add primary antibody in 0.2% D-PBS/BSA solution to the panning dishes as follows:
    i. Dish 1: 1 mL of RAN-2 supernatant in 4 mL of D-PBS/0.2% BSA.
    ii. Dish 2: 1 mL of GC supernatant in 4 mL of D-PBS/0.2% BSA.
    iii. Dish 3: A2B5 ascites 1:750 in D-PBS/0.2% BSA.
       A2B5 supernatant can also be used (at about 1:20). Immediately before adding the cell suspension to each dish, rinse with three washes of D-PBS. For mouse optic nerve OPCs, use O4 instead of A2B5 and Thy1.2 instead of RAN-2 (see Chapter 8, Protocol 1: Purification of Oligodendrocyte Precursor Cells from Rat Cortices by Immunopanning [Dugas and Emery]).
25. Incubate the dishes at room temperature for at least 1 h before use.
26. Equilibrate 10 mL of EBSS at 37°C (10% CO_2).
27. Prepare 30% FCS as follows.
    i. Mix 4.5 mL of FCS with 10.5 mL of D-PBS.
    ii. Filter through a 0.22-µm filter to sterilize.
    iii. Equilibrate to room temperature.
28. Dissect the optic nerves and optic chiasm from three litters of P7 Sprague–Dawley rat pups as follows.
    i. Remove the head and skull and lift the brain out.
       The optic nerves will be lying on the base of the skull.
    ii. Cut the optic nerves behind the chiasm and also behind each eye. Grasp the chiasm with forceps.
    iii. Transfer the pair of nerves into a 35-mm Petri dish containing 2 mL of D-PBS.
29. Cut each nerve with fine scissors into as many pieces as possible (at least five to 10 pieces per nerve). Transfer the D-PBS and nerve pieces (excluding chiasm pieces) to a small tube.
30. Prepare a papain solution as follows.
    i. Add 165 U of papain to 5 mL of D-PBS.
    ii. Add 100 µL of 0.4% DNase stock.
    iii. Dissolve the papain by placing this solution in a water bath at 37°C for a few minutes.
    iv. Just before use, add L-cysteine (2 mg) to activate the papain solution.
    v. Adjust the pH to 7.4 with 1 M NaOH.
    vi. Filter through a 0.22-µm filter to sterilize.

Cite this protocol as *Cold Spring Harb Protoc*; doi:10.1101/pdb.prot074971

31. Remove the D-PBS from the harvested nerve tissue with a Pasteur pipettes and add the papain solution.
32. Incubate the tissue, with occasional agitation, for 75 min in a water bath at 37°C.
33. Remove the papain solution with a Pasteur pipette.
34. Rinse away the excess papain by adding 3 mL of low-ovo solution. Let the tissue pieces settle and remove the solution.
35. Serially triturate the nerve pieces in low-ovo solution, 1 mL at a time, with a 2.5-mL syringe and #21 gauge needle (three sets of five passes each), followed by a #23 gauge needle (three sets of five passes each), as follows.
    i. Add 1 mL of low-ovo solution, pass the nerve pieces through the needle one or two times, let the pieces settle, and collect the suspension above the pieces with a Pasteur pipette into a 15-mL conical tube.
    ii. Add another 1 mL of low-ovo solution, triturate again, and repeat as described.

    With this method, cells successfully removed from the nerve pieces are not repetitively passed through the needle.
36. Filter the cell suspension through an autoclaved 20-μm nylon mesh as follows.
    i. Form a cone with the filter on top of a 15-mL tube.
    ii. Prewet the filter by adding 0.5 mL of low-ovo solution.
    iii. Transfer the cell suspension through the filter.
    iv. Add the remaining volume of low-ovo solution (~0.5 mL).
37. Centrifuge the dissociated cells at 200*g* (1000 rpm) in a clinical centrifuge for 13 min at 25°C. Aspirate the supernatant.
38. Resuspend the cells in 1 mL of high-ovo solution, then mix in the remaining 5 mL of high-ovo solution.
39. Centrifuge for 13 min at 25°C. Aspirate the supernatant.

    If transfection is required, proceed to the section Nucleofection of Optic Nerve Cells following Step 47. Complete Steps 48–58 before continuing OPC purification at Step 40. If no transfection is needed, proceed directly to Step 40–47 to finish OPC purification.
40. Resuspend the optic nerve cells in 8 mL of panning buffer.
41. Proceed with immunopanning as follows.
    i. Rinse each panning dish (from Step 25) three times with D-PBS immediately before adding the cell suspension.
    ii. Add the cell suspension to Dish 1 (Ran 2, negative selection for astrocytes).
    iii. Incubate for 30 min at room temperature, agitating gently at 15 min to ensure access of all cells to the panning surface area.
    iv. Transfer the cell suspension to Dish 2 (GC, negative selection for OLs).
    v. Incubate for 20–30 min at room temperature, agitating at 10–15 min.
    vi. Transfer the cell suspension to Dish 3 (A2B5, positive selection for OPCs).
    vii. Incubate for 30 min at room temperature, agitating at 15 min.
42. Wash Dish 3 (A2B5) with D-PBS (~10 mL per wash) around five times to remove nonadherent cells. Check the success of the washes under a microscope.
43. Release the purified cells from the dish as follows.
    i. Add 200 μL of 2.5% trypsin to 4 mL of preequilibrated EBSS.

ii. Rinse Dish 3 (A2B5) once with the remaining 6 mL of preequilibrated EBSS.
iii. Remove the EBSS wash and add the 4 mL of trypsin/EBSS.
iv. Place at 37°C (10% CO_2) for 10 min.
v. Add 2 mL of 30% FCS.
vi. Use a P1000 pipette to gently squirt around the surface of the dish to release the cells.
vii. Collect the supernatant in a 15-mL conical tube containing 1 mL of 30% FCS.
viii. Repeat Steps 43.vi and 43.vii twice, adding another 3–4 mL of 30% FCS to the dish each time.

44. Centrifuge the purified OPCs at 200*g* (1000 rpm) for 13 min.
45. Aspirate the supernatant carefully but completely.
46. Plate the purified OPCs onto the RGC reaggregate cultures as follows.

i. Resuspend the OPCs in ND-G medium or MyM (20,000 OPCs in 0.5 mL medium).
ii. For each well of RGC reaggregates, remove 300 µL of medium and add 500 µL of OPCs.

Cocultures may be maintained for up to 6 d with no further manipulation.

47. For culture periods >6 d, change one-half volume of medium every 3 d. When appropriate, add the γ-secretase inhibitor DAPT to a final concentration of 1 µM to promote ensheathment.

Differentiation and ensheathment occur well in either ND-G or MyM. MyM supports more reliable and robust wrapping and compaction over a short time course. MyM, however, does not support initial survival and recovery of acutely isolated RGCs as well as does ND-G. Replacement of ND-G (for purified RGC axon outgrowth) with MyM (for myelinating cocultures) is best achieved by beginning the switch with a half volume medium change in the final one or two RGC feedings before OPC isolation. The plating of OPCs as described here further reduces ND-G, and subsequent half volume feedings are sufficient to complete the changeover.

Nucleofection of Optic Nerve Cells

48. Prepare 5 mL of ND-G medium.
49. Prepare a PDL-coated tissue culture dish as follows.

i. Dilute 100 µL of PDL stock in 10 mL of sterile distilled H_2O.
ii. Incubate a 10-cm tissue culture dish with 5 mL of PDL solution for 30 min at room temperature.
iii. Rinse three times with D-PBS.

50. Make 10 mL of 20% FCS as follows.

i. Mix 2 mL of FCS with 8 mL of D-PBS.
ii. Filter through a 0.22-µm filter to sterilize.
iii. Equilibrate to room temperature.

51. Resuspend the optic nerve cells from Step 39 in 500 µL of ND-G medium.
52. Add all the cells to the PDL-coated dish containing the remaining (~4.5 mL) equilibrated ND-G medium.
53. Allow the cells to recover in the incubator for 90 min at 37°C.
54. Release the cells from the dish as follows.

i. Remove the medium and rinse the cells once with 6 mL of preequilibrated EBSS.
ii. Incubate the cells at 37°C with 4 mL of trypsin-EDTA (diluted 1:10 in EBSS) for 8 min.
iii. Remove the cells from the incubator and add 2 mL of 20% FCS.

Cite this protocol as *Cold Spring Harb Protoc*; doi:10.1101/pdb.prot074971

iv. Use a P1000 pipette to gently squirt around the surface of the dish to release the cells.
v. Collect the supernatant in a 15-mL conical tube containing 1 mL of 20% FCS.
vi. Repeat Steps 54.iv and 54.v twice, adding another 3–4 mL of 20% FCS each time.

55. Centrifuge the cells at 200*g* (1000 rpm) for 11 min.
56. Resuspend the cells in Lonza nucleofector solution. Use 105 µL of solution per transfection.
57. Perform nucleofection for each of the desired transfections as follows.
    i. Quickly transfer 100 µL of the cell suspension to a tube containing 1.5 µg of plasmid or siRNA.
    ii. Mix gently.
    iii. Transfer this volume to the provided cuvette.
    iv. Electroporate the cells using program O-17.
    v. Add 500 µL of ND-G to the cuvette and use the provided plastic pipette to transfer the entire volume of cells to a tube containing 8 ml of panning buffer (see Step 40).
58. Finish the isolation of optic nerve OPCs by proceeding with Steps 40–47.

RELATED INFORMATION

This protocol was adapted from Watkins et al. (2008), © 2008, with permission from Elsevier.

RECIPES

Biotin Stock (5000×)

Dissolve 10 mg of biotin (Sigma-Aldrich B4639) in 200 mL of D-PBS with phenol red to prepare biotin stock. Add 0.1 N NaOH in 5-µL drops to aid in dissolving. Make 10-µL aliquots and store at −20°C.

BSA Stock (4%)

Dissolve 2 g of BSA (Sigma-Aldrich A4161) in 50 mL of D-PBS at 37°C. Adjust pH to 7.4 with ~200 µL of 1 N NaOH. Filter sequentially through 0.45-µm and 0.22-µm filters. Prepare 1.0-mL aliquots and store at −20°C.

CNTF Stock (10 µg/mL)

To prepare, dilute ciliary neurotrophic factor (CNTF; Peprotech 450-02) to 10 µg/mL with sterile 0.2% BSA in Dulbecco's phosphate-buffered saline (D-PBS; HyClone SH30264.01). Make 20-µL aliquots, flash freeze in liquid nitrogen, and store –80°C.

Forskolin Stock (4.2 mg/mL)

To prepare, add 1 mL of sterile DMSO to a 50-mg bottle of forskolin (Sigma-Aldrich F6886) and pipette up and down until the powder is fully resuspended. Transfer to a 15-mL conical tube and add an additional 11 mL of DMSO to achieve a final concentration of 4.2 mg/mL. Store in 20- and 80-µL aliquots at −20°C.

High-Ovomucoid Stock (6×)

1. Dissolve the following in 160–180 mL of Dulbecco's phosphate-buffered saline (D-PBS; Invitrogen 14287-080).

Reagent	Amount (for 200 mL)	Final concentration
BSA (Sigma-Aldrich A-8806)	6 g	30 mg/mL
Trypsin inhibitor (Worthington LS003086)	6 g	30 mg/mL

2. Adjust the pH to 7.4 with 10 N NaOH. Bring the volume to 200 mL with D-PBS, then filter-sterilize.
3. Make 1-mL aliquots, and store them at −20°C.

Hormone Mix (200×)

Reagents	Amount to add	Final concentration
Apotransferrin (Sigma-Aldrich T1147; 4 mg/mL in DMEM)	5 mL	1 mg/mL
Putrescine (Sigma-Aldrich P5780; 12.8 mg/mL in DMEM)	5 mL	20 mM
Progesterone (Sigma-Aldrich P8783; 25 μg/μL in ethanol)	10 μL	4 μM
Sodium selenite (Sigma-Aldrich S5261; 0.4 mg/mL in DMEM)	51.9 μL	6 μM

Combine the ingredients in an additional 10 mL of DMEM. Mix well and filter through a 0.22-μm filter. Make 200-μL aliquots and store at −20°C.

Insulin Stock (0.5 mg/mL)

To 20 mL of sterile water, add 10 mg of insulin (Sigma-Aldrich I6634) and 100 μL of 1.0 N HCl. Mix well. Filter through a 0.22-μm filter. Store at 4°C for 4–6 wk.

Low-Ovomucoid Stock Solution (10×)

To prepare, add 3 g of BSA (Sigma-Aldrich A8806) to 150 mL D-PBS. Mix well. Add 3 g of trypsin inhibitor (Worthington LS003086) and mix to dissolve. Add ~1 mL of 1 N NaOH to adjust the pH to 7.4. Bring the volume to 200 mL with D-PBS. Filter-sterilize through a 0.22-μm filter. Make 1.0-mL aliquots and store at −20°C.

Myelination Medium (MyM)

1. Dilute a 10 μg/mL stock solution of CNTF to 0.1 ng/mL by combining the following:

   1 μL CNTF stock (10 μg/mL) <R>

   99 μL MyM base medium <R>

2. Add 15 μL of 0.1 ng/mL CNTF to the following:

   15 mL MyM base medium <R>

   15 μL BDNF (Peprotech 450-02; 50 μg/mL in D-PBS [Gibco 14287-080] containing 0.2% BSA)

3. Equilibrate in an incubator to 37°C/10% CO_2.

Cite this protocol as *Cold Spring Harb Protoc*; doi:10.1101/pdb.prot074971

MyM Base Medium

Reagent	Amount to add (for 40 mL)
DMEM-High Glucose medium (Gibco 11960-044)	39 mL
Insulin stock (0.5 mg/mL) <R>	400 µL
Sodium pyruvate (100 mM) (Gibco 11360-070)	400 µL
L-Glutamine (200 mM) (Gibco 25030-081)	400 µL
Hormone mix (200×) <R>	200 µL
T3 stock (4 µg/mL) <R>	200 µL
Hydrocortisone (50 µM) (Sigma-Aldrich H6909)	40 µL
Trace Elements B (1000×) (Cellgro)	40 µL
Biotin stock (5000×) <R>	8 µL
Vitamin B12 (Sigma-Aldrich V6629; 1.36 mg/mL in D-PBS)	8 µL
B-27 (Invitrogen) or N21-MAX (R&D Systems)	15 µL
Ceruloplasmin (1 mg/mL) (Calbiochem 239799)	4 µL

Filter the solution through a 0.22-µm filter to sterilize and store at 4°C for up to 1 wk.

NAC Stock (5 mg/mL)

To prepare, dissolve 50 mg of *N*-acetyl-L-cysteine (NAC) powder (Sigma-Aldrich A8199) in 10 mL of Neurobasal Medium (Gibco/Life Technologies 21103). (The solution will be yellowish.) Filter through a 0.22-µm filter. Prepare 20- and 80-µL aliquots and store them at −20°C.

ND-Growth Medium (ND-G)

Reagent	Amount to add
ND-SATO base medium <R>	8 mL
BDNF (Peprotech 450-02; 50 µg/mL in D-PBS containing 0.2% BSA)	8 µL
CNTF (10 µg/mL) <R>	8 µL
Forskolin stock (4.2 mg/mL) <R>	8 µL

Equilibrate in an incubator to 37°C/10% CO_2.

ND-SATO Base Medium

Reagent	Amount to add (for 40 mL)
Neurobasal solution (Gibco 21103-049)	19.5 mL
DMEM (Gibco 11960-044)	19.5 mL
Penicillin/streptomycin (100×) (Gibco 15140-122)	400 µL
L-Glutamine (200 mM) (Gibco 25030-081)	400 µL
SATO (100×) <R>	400 µL
Sodium pyruvate (100 mM) (Gibco 11360-070)	400 µL
Insulin stock (0.5 mg/mL) <R>	400 µL
T3 stock (4 µg/mL) <R>	400 µL
NAC stock (5 mg/mL) <R>	40 µL
Trace Elements B (1000×) (Cellgro)	40 µL
Biotin stock (5000×) <R>	8 µL
B-27 (Invitrogen) or N21-MAX (R&D Systems)	800 µL

Filter the solution through a 0.22-µm filter to sterilize and store at 4°C for up to 1 wk.

SATO (100×)

1. Prepare the following stock solutions (these should be made fresh; do not reuse).
   - Combine 2.5 mg of progesterone (Sigma-Aldrich P8783) and 100 µL of ethanol to make a progesterone stock solution.
   - Combine 4.0 mg of sodium selenite (Sigma-Aldrich S5261), 10 µL of 1 N NaOH, and 10 mL of Neurobasal (NB, Gibco 21103-049) to make a sodium selenite stock solution.
2. Add the following to 40 mL of DMEM (Gibco 11960-044):

Reagent	Quantity	Final concentration in medium (1×)
BSA (Sigma-Aldrich A4161)	400 mg	100 µg/mL
Transferrin (Sigma-Aldrich T1147)	400 mg	100 µg/mL
Putrescine dihydrochloride (Sigma-Aldrich P5780)	64 mg	16 µg/mL
Progesterone stock solution	10 µL	60 ng/mL
Sodium selenite stock solution	400 µL	40 ng/mL

3. Mix well, and filter-sterilize through a prerinsed 0.22-µm filter. Make 200-µL aliquots, and store at −20°C.

T3 Stock (4 µg/mL)

Dissolve 3.2 mg of 3,3′,5-triiodo-L-thyronine sodium salt (T3; Sigma-Aldrich T6397) in 400 µL of 0.1 N NaOH. Add 10 µL of T3 solution to 20 mL of Dulbecco's phosphate-buffered saline (D-PBS; Gibco 14190-144). Filter through a 0.22-µm filter, discarding the first 10 mL. Make 200-µL aliquots, and store at −20°C.

REFERENCES

Dugas JC, Emery B. 2013. Purification of oligodendrocyte precursor cells from rat cortices by immunopanning. *Cold Spring Harb Protoc* doi: 10.1101/pdb.prot070862.

Watkins TA, Emery B, Mulinyawe S, Barres BA. 2008. Distinct stages of myelination regulated by γ-secretase and astrocytes in a rapidly myelinating CNS coculture system. *Neuron* **60:** 555–569.

Winzeler A, Wang JT. 2013. Purification and culture of retinal ganglion cells from rodents. *Cold Spring Harb Protoc* doi: 10.1101/pdb.prot074906.

Cite this protocol as *Cold Spring Harb Protoc;* doi:10.1101/pdb.prot074971

CHAPTER 10

Purification of Schwann Cells

Amanda Brosius Lutz[1]

Stanford University School of Medicine, Department of Neurobiology, Stanford, California 94305

This chapter introduces methods for the acute purification and primary culture of Schwann cells from the mouse sciatic nerve. Immunopanning can be used to isolate Schwann cells from intact nerves during early postnatal development as well as to purify Schwann cells from adult nerves following sciatic nerve injury. These methods facilitate the exploration of mouse Schwann cell biology in the healthy and injured peripheral nerve.

ISOLATING SCHWANN CELLS

Schwann cells are the principal glial cell type of the peripheral nervous system (PNS) and play key roles in peripheral nerve development, function, and regeneration after injury (Jessen and Mirsky 1999; Stevens and Fields 2000; Chen et al. 2007). Despite recent advances, many questions remain regarding mechanisms of Schwann cell function and dysfunction in the healthy and diseased PNS.

The ability to purify glia and study them in vitro alone or in coculture with other cell types has proved a valuable tool for advancing our understanding of nervous system function. Isolation of a pure Schwann cell population presents two main challenges. First, the peripheral nerve tissue must be dissociated into single cells, and second, Schwann cells must be separated from macrophages, fibroblasts, and axon and myelin debris. The challenge of tissue dissociation becomes significantly more difficult as the nerve matures and acquires extensive myelin and more substantial amounts of connective tissue (Bunge et al. 1989). In early postnatal tissue, successful dissociation can be accomplished enzymatically. In the adult intact nerve, however, enzymatic dissociation yields very low cell numbers and viability (Morrissey et al. 1991). This hurdle can be overcome by including a predegeneration stage, either in vivo or in situ, allowing time for Wallerian degeneration and Schwann cell proliferation (Brockes et al. 1979; Salzer and Bunge 1980; Morrissey et al. 1991; Fernandez-Valle et al. 1995; Komiyama et al. 2003; Mauritz et al. 2004).

Once nerve dissociation is complete and a single cell suspension is achieved, Schwann cell purification can be carried out using a variety of techniques. Reported methods include the use of antimitotic agents such as cytosine arabinoside (AraC) to selectively limit fibroblast growth (Brockes et al. 1979), complement-mediated lysis of fibroblasts using antibodies to Thy-1 (Brockes et al. 1979; Needham et al. 1987; Morrissey et al. 1991), antibody-labeled magnetic beads for depletion or selection of cells based on target molecules, such as the p75 neurotrophin receptor on the surface of Schwann cells (Vroemen and Weidner 2003; Haastert et al. 2007), a "cold jet" technique, which targets differential cell-substrate adhesion strength between fibroblasts and Schwann cells (Jirsova et al. 1997; Teare et al. 2004), repeated replating of nerve explants (Askanas et al. 1980; Morrissey et al. 1991), and, most recently, selection of Schwann cells based on preferential capacity for metabolism of D-valine (Kaewkhaw et al. 2012). Immunopanning, through its sequence of negative and positive

[1]Correspondence: abrolutz@stanford.edu

Cite this introduction as *Cold Spring Harb Protoc;* doi:10.1101/pdb.top073981

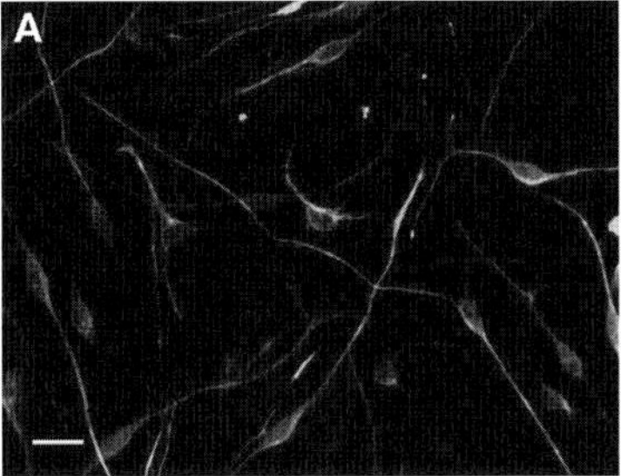

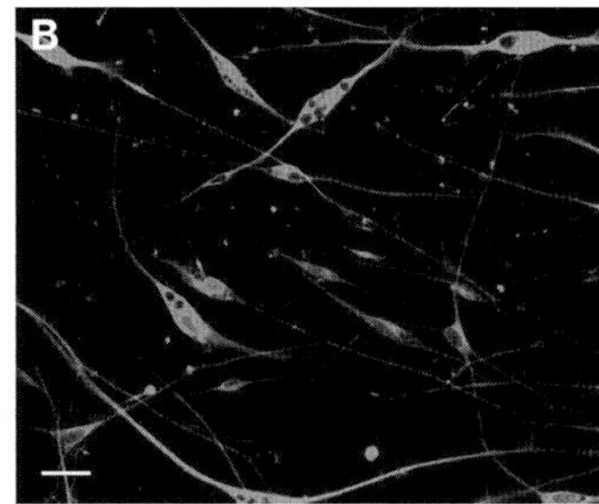

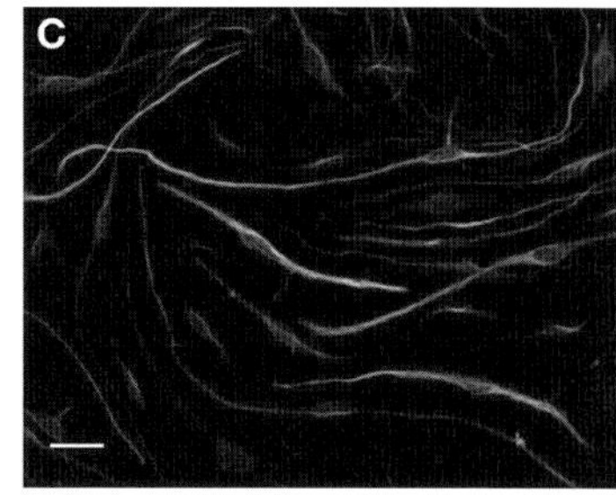

FIGURE 1. (*A*) Neonatal mouse Schwann cells isolated at P2 after 3 d in vitro. Schwann cells are labeled with S100. (*B*) Adult mouse Schwann cells isolated 6 d postsciatic nerve crush at 1 d in vitro. Schwann cells are labeled with S100. Note that Schwann cells are actively phagocytosing myelin debris. (*C*) Neonatal rat Schwann cells isolated at P2 after 3 d in vitro. Schwann cells are labeled with S100. Scale bars, 20 μm.

selection steps, has also been used successfully to separate neonatal Schwann cells from contaminating fibroblasts as well as macrophages (Cheng and Mudge 1996). Immunopanning is a gentle technique performed in ~1 h at room temperature that yields highly pure and viable cells (Barres et al. 1992).

CULTURING SCHWANN CELLS

The vast majority of protocols for Schwann cell culture to date describe methods for the culture of rat Schwann cells rather than mouse. Mouse Schwann cells are notoriously more difficult to culture, but the ability to successfully purify and grow these cells enables a broad range of experiments using transgenic mouse models. A protocol for the purification and culture of neonatal and adult mouse Schwann cells is described in Protocol 1: Purification of Schwann Cells from the Neonatal and Injured Adult Mouse Peripheral Nerve (Lutz). Similarly to Cheng and Mudge (1996), the described method utilizes immunopanning to deplete macrophages and fibroblasts from the nerve cell suspension and to positively select for Schwann cells. Using this method, it is possible to achieve >98% purity in neonatal (Fig. 1A) and adult (Fig. 1B) Schwann cell cultures. In our protocol for culture of mouse Schwann cells from the adult nerve, Wallerian degeneration proceeds in vivo after nerve crush and before animal euthanasia. Note that this protocol can be adopted for purification and culture of rat Schwann cells (Fig. 1C) with only minor modifications, as described.

The Schwann cells purified according to this protocol grow healthily at high density in a serum-free defined medium. The medium is a modification of the medium of Bottenstein and Sato (1979). At low cell density, the medium may need to be supplemented with PDGF-BB, IGF, and NT-3 to compensate for lack of autocrine factors (Meier et al. 1999). Cultures will proliferate in the presence of fibroblast growth factors and neuregulin (Cheng et al. 1998). For studies aimed at investigating the Schwann cell response to injury, 5%–10% serum can be added to the culture medium. Serum addition closely recapitulates the in vivo environment following injury-induced breakdown of the blood–nerve barrier and will sustain down-regulation of Schwann cell myelin proteins (Cheng and Mudge 1996; Vargas and Barres 2007).

ACKNOWLEDGMENTS

Many thanks to Lu Zhou for technical assistance and insight, Michio Painter for help with the protocol for dissociation, and Brad Zuchero for advice on cell culture conditions. This work was performed in Ben A. Barres' laboratory at Stanford University. His work was supported by the NIH (National Institutes of Health) R01 EY11310 (B.A.B.), the Adelson Medical Foundation (B.A.B), and the Christopher & Dana Reeve Foundation (B.A.B).

Cite this introduction as *Cold Spring Harb Protoc*; doi:10.1101/pdb.top073981

REFERENCES

Askanas V, Enge WK, Daiakas MC, Lawrence JV, Carter LS. 1980. Human Schwann cells in tissue culture: Histochemical and ultrastructural studies. *Arch Neurol* **37:** 329–337.

Barres BA, Hart IK, Coles HS, Burne JF, Voyvodic JT, Richardson WD, Raff MC. 1992. Cell death and control of cell survival in the oligodendrocyte lineage. *Cell* **70:** 31–46.

Bottenstein JE, Sato GH. 1979. Growth of a rat neuroblastoma cell line in serum-free supplemented medium. *Proc Natl Acad Sci* **76:** 514–517.

Brockes JP, Fields KL, Raff MC. 1979. Studies on cultured rat Schwann cells I. Establishment of purified populations from cultures of peripheral nerve. *Brain Res* **165:** 105–118.

Bunge MB, Wood PM, Tynan LB, Bates ML, Sanes JR. 1989. Perineurium originates from fibroblasts: Demonstration in vitro with a retroviral marker. *Science* **243:** 23–229.

Chen ZL, Yu WM, Strickland S. 2007. Peripheral regeneration. *Annu Rev Neurosci* **30:** 209–233.

Cheng L, Mudge AW. 1996. Cultured Schwann cells constitutively express the myelin protein P0. *Neuron* **16:** 309–319.

Cheng L, Esch FS, Marchionni MA, Mudge AW. 1998. Control of Schwann cell survival and proliferation: Autocrine factors and neuregulins. *Mol Cell Neurosci* **2:** 141–156.

Fernandez-Valle C, Bunge RP, Bunge MB. 1995. Schwann cells degrade myelin and proliferate in the absence of macrophages: Evidence from in vitro studies of Wallerian degeneration. *J Neurocytol* **24:** 667–679.

Haastert K, Mauritz C, Chaturvedi S, Grothe C. 2007. Human and rat adult Schwann cell cultures: Fast and efficient enrichment and highly effective non-viral transfection protocol. *Nat Protoc* **2:** 99–104.

Jessen KR, Mirsky R. 1999. Schwann cells and their precursors emerge as major regulators of nerve development. *Trends Neurosci* **22:** 402–410.

Jirsova K, Sodaar P, Mandys V, Bar PR. 1997. Cold jet: A method to obtain pure Schwann cell cultures without the need for cytotoxic, apoptosis inducing drug treatment. *J Neurosci Meth* **78:** 133–137.

Kaewkhaw R, Scutt AM, Haycock JW. 2012. Integrated culture and purification of rat Schwann cells from freshly isolated adult tissue. *Nat Protoc* **7:** 1996–2004.

Komiyama T, Nakao Y, Toyama Y, Asou H, Vacanti CA, Vacanti MP. 2003. A novel technique to isolate adult Schwann cells for an artificial nerve conduit. *J Neurosci Meth* **122:** 195–200.

Lutz A. 2014. Purification of Schwann cells from the neonatal and injured adult mouse peripheral nerve. *Cold Spring Harb Protoc* doi: 10.1101/pdb.prot074989.

Mauritz C, Grothe C, Haastert K. 2004. Comparative study of cell culture and purification methods to obtain highly enriched cultures of proliferating adult rat Schwann cells. *J Neurosci Res* **77:** 453–461.

Meier C, Parmantier E, Brennan A, Mirsky R, Jessen KR. 1999. Developing Schwann cells acquire the ability to survive without axons by establishing an autocrine circuit involving insulin-like growth factor, neurotrophin-3, and platelet-derived growth factor-BB. *J Neurosci* **19:** 3847–3859.

Morrissey TK, Kleitman N, Bunge RP. 1991. Isolation and functional characterization of Schwann cells derived from adult peripheral nerve. *J Neurosci* **11:** 2433–2442.

Needham LK, Tennekoon GI, McKhann GM. 1987. Selective growth of rat Schwann-cells in neuron-free and serum-free primary culture. *J Neurosci* **7:** 1–9.

Salzer JL, Bunge RP. 1980. Studies of Schwann cell proliferation. I. An analysis in tissue culture of proliferation in development, Wallerian degeneration, and direct injury. *J Cell Biol* **84:** 739–752.

Stevens B, Fields RD. 2000. Response of Schwann cells to action potentials in development. *Science* **287:** 2267–2271.

Teare KA, Pearson RG, Shakesheff KM, Haycock JW. 2004. α-MSH inhibits inflammatory signalling in Schwann cells. *Neuroreport* **15:** 493–498.

Vargas ME, Barres BA. 2007. Why is Wallerian degeneration in the CNS so slow? *Annu Rev Neurosci* **30:** 153–179.

Vroemen M, Weidner N. 2003. Purification of Schwann cells by selection of p75 low affinity nerve growth factor receptor expressing cells from adult peripheral nerve. 2003. *J Neurosci Meth* **124:** 135–143.

Protocol 1

Purification of Schwann Cells from the Neonatal and Injured Adult Mouse Peripheral Nerve

Amanda Brosius Lutz[1]

Stanford University School of Medicine, Department of Neurobiology, Stanford, California 94305

This protocol describes the use of immunopanning for acute purification and primary culture of Schwann cells from intact neonatal and injured adult mouse sciatic nerve.

MATERIALS

It is essential that you consult the appropriate Material Safety Data Sheets and your institution's Environmental Health and Safety Office for proper handling of equipment and hazardous material used in this protocol.

RECIPES: Please see the end of this protocol for recipes indicated by <R>. Additional recipes can be found online at http://cshprotocols.cshlp.org/site/recipes.

Reagents

Bovine serum albumen (BSA) stock (4%) <R>

Collagenase/dispase cocktail <R>

DNase

To prepare a 0.4% stock of DNase in Earle's balanced salt solution (EBSS), add 1 mL of EBSS (Sigma-Aldrich E6267) per 12,500 units of DNase I (Worthington LS002007). Keep on ice. Filter-sterilize, and store in 200-μL aliquots at −30°C.

Dulbecco's modified Eagle medium (DMEM) (Invitrogen 11960-044)

Dulbecco's phosphate-buffered saline (D-PBS) (Gibco 14287)

Earle's balanced salt solution (EBSS) (Sigma-Aldrich E6267)

Fetal calf serum (FCS) (Gibco 10437-028)

Prepare 50-mL aliquots and heat inactivate for 30 min at 55°C, and then store at −20°C.

Immunopanning reagents

- Mouse anti-mouse Thy1.2 antibody (AbD Serotec MCA02R)

  If isolating rat Schwann cells, use T11D7 Thy-1.1 hybridoma (mouse IgM) instead of Thy1.2 antibody.

- O4 hybridoma supernatant (mouse IgM) (Bansal et al. 1989)
- Rat anti-mouse CD45 antibody (BD Pharmingen 550539)
- Goat anti-mouse IgG + IgM (H + L) secondary antibody (Jackson ImmunoResearch 115-005-044)
- Goat anti-mouse IgM, μ-chain specific secondary antibody (Jackson ImmunoResearch 115-005-020)
- Goat anti-rat IgG (H + L) secondary antibody (Jackson ImmunoResearch 112-005-167)

Insulin stock (0.5 mg/mL) <R>

[1]Correspondence: abrolutz@stanford.edu

Cite this protocol as *Cold Spring Harb Protoc*; doi:10.1101/pdb.prot074989

Laminin (mouse) (Cultrex; R&D Systems 3400-010-01)

Thaw mouse laminin stock (1 mg/mL) at 4°C. Make 10-µL aliquots and store at −30°C.

Low-ovomucoid (low-ovo) stock solution (10×) <R>

Just before use (Step 15), prepare a low-ovo working solution by combining 9 mL of D-PBS, 1 mL of 10× low-ovo stock solution, and 100 µL of 0.4% DNase.

Mice (neonatal or adult)

For neonatal Schwann cells, use four or five litters of P2 mice (30 pups). For postinjury adult Schwann cells, use four mice at 5–6 d after nerve crush injury and only collect tissue distal to the site of injury. Schwann cells have been isolated successfully from injured nerves of mice up to 24 mo of age. Rats can be substituted for mice.

Panning buffer

Just before use (Step 20), prepare panning buffer by combining 9 mL of D-PBS, 1 mL of 0.2% BSA in D-PBS, and 100 µL of insulin stock solution.

Poly-D-lysine

Add 5 mL of H_2O to a 5-mg bottle of poly-D-lysine (pDL; Sigma-Aldrich P6407). Filter through a 2-µm filter. Make 100-µL aliquots and store at −20°C. Dilute stock to 20 µg/mL in sterile H_2O before use.

Schwann cell growth medium <R>

Tris-HCl (50 mM, pH 9.5) (sterile)

Dissolve 12.1 g of Trizma base in 200 mL of dH_2O. Adjust pH to 9.5 with HCl.

Trypsin (2.5%) (Invitrogen 15090–046)

Equipment

Centrifuge (table-top clinical, with 15-mL tube adaptors)

Dissection equipment

- Decapitation scissors
- Forceps (#5), two pairs
- Microdissection scissors

Ethanol-washed glass coverslips <R>

Tissue culture dishes (10-cm; Falcon or Nunc) may be substituted for coverslips when plating cells.

Forceps

Hemocytometer

Nitex mesh filter (3–20/14; Tetko)

Cut Nitex mesh into 3 inch squares, wrap in small packets of foil, and autoclave.
Spray with ethanol and use Bunsen flame to sterilize.

Pasteur pipettes (glass)

No fire polishing is necessary.

Petri dishes (10 cm, for panning dishes) (Falcon or Nunc)

Syringe filters (0.22 µm)

Tissue culture dish (35 mm)

Tissue culture hood (sterile)

Tissue culture incubator (10% CO_2, 37°C)

Tissue culture plates (multiwell)

Tubes (15-mL conical)

METHOD

An outline of Schwann cell purification is provided in Figure 1. Perform Steps 1–3 on the first day (~1 h) and the remaining steps on the following day.

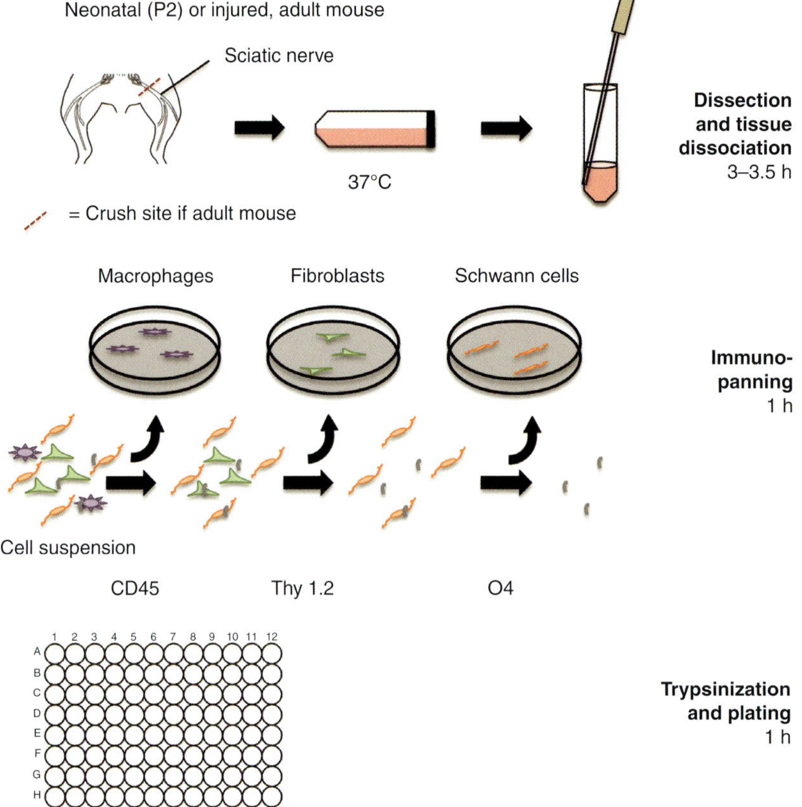

FIGURE 1. Flow diagram of protocol for purifying mouse Schwann cells.

Preparation

1. Coat coverslips with poly-D-lysine as follows.
   i. In a sterile hood, rinse ethanol-washed glass coverslips four times with sterile dH_2O. Place one coverslip per well in a multiwell tissue culture plate.
   ii. Add poly-D-lysine (20 µg/mL) to each well and incubate for 30 min at room temperature. Ensure that the coverslips are not floating and are completely submerged.
   iii. Discard the poly-D-lysine solution and rinse each well three times with sterile dH_2O.
   iv. Dilute the mouse laminin stock 1:500 in 80% DMEM/20% dH_2O. Add this solution to each well so that the coverslips are submerged.

      Alternatively, use tissue culture dishes instead of coverslips and coat as described.

2. Prepare the panning plates by coating three 10-cm Petri dishes with secondary antibodies. For each dish, add 30 µL of secondary antibody in 10 mL of 50 mM Tris-HCl as follows:
   i. Dish 1: goat anti-rat IgG (for CD45).
   ii. Dish 2: goat anti-mouse IgG + IgM (for Thy-1.2).
   iii. Dish 3: goat anti-mouse IgM, µ chain specific (for O4).
3. Swirl the plates so that the bottom surface of each plate is completely covered and incubate the dishes at 4°C overnight.

   Plates are initially hydrophobic, but after coating overnight, plates become visibly hydrophilic. If plates are needed immediately, a quick but not ideal way of preparing plates is to coat with secondary antibodies for 2 h at 37°C. For best results, however, coat the plates with secondary antibodies overnight at 4°C.

4. Rinse the panning dishes three times with D-PBS.

 Cite this protocol as *Cold Spring Harb Protoc*; doi:10.1101/pdb.prot074989

5. Block nonspecific binding on the panning dishes by adding a solution of 0.2% BSA in D-PBS to each dish and incubate at room temperature for 30 min.
6. Remove the BSA and coat the panning dishes with primary antibodies as follows.
    i. For Dish 1 (CD45), add 10 µL of rat anti-mouse CD45 in 6 mL of 0.2% BSA in D-PBS.
    ii. For Dish 2 (Thy1.2), add 10 µL of mouse anti-mouse Thy1.2 in 6 mL of 0.02% BSA in D-PBS. (If isolating rat Schwann cells, use 6 mL of undiluted T11D7 hybridoma [i.e., Thy-1.1 instead of Thy-1.2].)
    iii. For Dish 3 (O4), add 1 mL of O4 hybridoma in 5 mL of 0.2% BSA in D-PBS.
7. Incubate the dishes at room temperature for a minimum of 2 h.
8. Rinse the dishes three times with D-PBS.

    The dishes are now ready for immunopanning.

Dissection and Dissociation

9. Dissect the sciatic nerves from each mouse as follows.
    i. Cut through the skin around the back and abdominal region. Remove the skin over the hind limbs and position the mouse ventral side down with hind legs stretched out.
    ii. Make an incision parallel to the spine just adjacent to the vertebral column mid-thigh and locate the sciatic nerve. Slice open the dorsal thigh muscle to reveal the sciatic nerve from the hip to the knee.
    iii. Collect the nerves in a 35-mm dish containing D-PBS. If dissecting injured nerves, only collect the segment of nerve distal to the crush site.
10. Transfer the nerves into a 15-mL conical tube and centrifuge at 1000 rpm (220*g*) for 3 min to pellet the nerve tissue.
11. Aspirate the D-PBS from the nerve tissue.
12. Add 3 mL of warmed collagenase/dispase (C/D) enzyme cocktail to the nerve tissue and incubate the tube horizontally at 37°C for 1.5 h.
13. During enzymatic dissociation (Step 12), preequilibrate 10 mL each of EBSS and Schwann cell growth medium in an incubator (10% CO_2, 37°C).
14. Centrifuge the nerve tissue at 220*g* for 3 min.
15. Aspirate the enzyme and replace the solution with 3 mL of low-ovo working solution.
16. Dissociate the nerve tissue mechanically with a glass Pasteur pipette by triturating vigorously 30–40 times or until only small wisps of tissue remain visible. Avoid introducing bubbles into the solution, because this will diminish cell yield.

    See Troubleshooting.
17. Centrifuge the cell suspension at 220*g* for 10 min.

    If Schwann cells are purified from the injured sciatic nerve 4–6 d after crush injury, the cell suspension will contain a substantial amount of myelin debris. Small debris can be reduced by centrifuging cells through a BSA cushion: Layer the cell suspension gently over 3 mL of 4% BSA and centrifuge without using the brake. To remove large pieces of myelin debris, use a three-layered filter in Step 19.
18. Resuspend the cell pellet in 3 mL of panning buffer.
19. Make a Nitex filter cone and use flamed forceps to hold the filter in place. Filter the cell solution 1 mL at a time through the mesh to remove any remaining clumps of cells and chunks of tissue.
20. Wash the filter with 3 mL of panning buffer.
21. Bring the volume up to 6 mL with panning buffer. Count the cells using a hemocytometer, if desired.

Immunopanning

When transferring cells between panning plates (Steps 22–24), extra panning buffer can be used to rinse the surface of the previous plate to wash off any nonadherent cells left behind after pouring. However, one should take care not to increase the total volume of cell suspension above 7–8 mL.

22. Pour the cells into the CD45 panning plate and incubate at room temperature for 20 min.

    See Troubleshooting.

23. Shake the plate and transfer the cells to the Thy1.2 plate. Incubate at room temperature for 20 min.

    See Troubleshooting.

24. Shake the plate and transfer the cells to the O4 plate. Incubate at room temperature for 20 min.

    See Troubleshooting.

25. During panning, prepare and filter 10 mL of 30% FCS in D-PBS.

Trypsinization and Plating

26. Wash the positive panning dish ~8× with D-PBS.
27. Perform one final rinse with 5 mL of equilibrated EBSS.
28. Add 200 μL of 2.5% trypsin to 4 mL of equilibrated EBSS. Pour this solution into the panning dish and incubate the dish for 5–7 min.
29. Tap the side of the dish to dislodge the cells. If no cells detach, return the dish to the incubator for one additional minute. If the cells come off easily with tapping, proceed to Step 30.

    See Troubleshooting.

30. Squirt the cells off the plate with 30% FCS as follows.
    i. Systematically squirt around the plate with a 1-mL pipette tip to dislodge the cells.
    ii. Transfer the dislodged cells to a 15-mL conical tube.
    iii. Examine the plate to ensure that no spots have been missed.
    iv. Remove 50 μL of cells for counting and centrifuge the remaining cells for 10 min at 220*g*.
    v. Count the cells using a hemocytometer.

       Expected cell yields are 10,000 cells per P2 mouse sciatic nerve and 50,000 cells per adult (8- to 10-week-old) mouse sciatic nerve (crushed).

31. Aspirate the solution from the cell pellet and resuspend the cells in equilibrated Schwann cell growth medium for preplating.

    See note at Step 33.

32. Just before plating, aspirate the laminin from the coverslips prepared in Step 1.iv.
33. Plate the cells in just enough volume to wet the surface of the coverslip (or the well of the culture dish).

    For 12-mm coverslips, we use 40 μL of cell suspension. For one well of a six-well plate, we use 100 μL and spread it evenly over the surface of the well with a fire-bent Pasteur pipette. This method of "preplating" ensures that a maximum number of buoyant Schwann cells settle onto the substrate and adhere to it rather than floating in a larger volume of medium. We plate cells at 10,000–12,000 cells/cm^2. This Schwann cell density ensures autocrine support of cell survival.

    See Troubleshooting.

34. Place plates in the incubator for 20 min to allow the cells to adhere to the substrate.
35. Add culture medium gently to the desired volume.
36. Feed the cells by replacing 50% of the medium with fresh Schwann cell growth medium every 5–7 d.

    For more information about culturing these cells, see Introduction: Purification of Schwann Cells (Lutz).

 Cite this protocol as *Cold Spring Harb Protoc*; doi:10.1101/pdb.prot074989

TROUBLESHOOTING

Problem (Step 16): Sciatic nerves are not dissociating upon mechanical trituration.
Solution: Cutting the nerves into smaller segments before incubation with enzymes and increasing the enzyme incubation time may increase the rate and uniformity of dissociation upon trituration.

Problem (Steps 22–24): Many clumps of cells are visible during Schwann cell panning.
Solution: Clumps of cells reduce yields and decrease cell purity. If clumps are visible, one can add a second Nitex filtration step just before O4 positive panning. One can also fire-polish the tip of the glass pipette used for trituration, although we find this to be unnecessary.

Problem (Step 24): Schwann cells seem to be floating during the positive panning step.
Solution: Reducing the volume of liquid in the positive panning plate can help to facilitate binding of lipid-rich, buoyant Schwann cells to the plate.

Problem (Step 29): Schwann cells are stuck too tightly to the O4 plate.
Solution: Depending on the batch of O4 hybridoma used, one may need to alter the dilution of the antibody to optimize cell adhesion during panning. As a rule of thumb, an optimal balance between positive cell selection and ease of trypsinization is achieved if 50% of Schwann cells are phase dark and 50% of cells are phase light at the end of panning on the O4 plate.

Problem (Step 33): Schwann cell cultures are littered with myelin debris.
Solution: Myelin debris is abundant in the injured nerve and binds to O4 antibodies. If this debris is not desired in final Schwann cell cultures, it can be rinsed off the coverslip/plate after cells have adhered to the surface by simply exchanging the culture medium. Alternatively, a triple Nitex filter and a BSA cushion can be used before immunopanning to remove myelin debris (see Step 17).

RECIPES

BSA Stock (4%)

Dissolve 2 g of BSA (Sigma-Aldrich A4161) in 50 mL of D-PBS at 37°C. Adjust pH to 7.4 with ~200 µL of 1 N NaOH. Filter sequentially through 0.45-µm and 0.22-µm filters. Prepare 1.0-mL aliquots and store at −20°C.

Collagenase/Dispase Cocktail

Dissolve 1 g of dispase II (Roche 04942078001) in 342 mL of 1× HEPES buffer. Remove and discard 42 mL of this solution. Dissolve 300 mg of collagenase A (Roche 10103578001) in the remaining 300 mL of dispase II solution. Filter to sterilize. Make aliquots of 10 mL and flash freeze. Store at −30°C.

Ethanol-Washed Glass Coverslips

Extensively wash 12-mm glass coverslips (Carolina Biological Supply 633029) in 70% ethanol. Perform the washes on a platform shaker in a beaker, with enough motion to lightly agitate the coverslips but not break too many. Wash the coverslips for about 1 mo, exchanging the ethanol approximately every day. (It is fine to skip some exchanges.) Store the washed coverslips in 70% ethanol until use.

Forskolin Stock (4.2 mg/mL)

To prepare, add 1 mL of sterile DMSO to a 50-mg bottle of forskolin (Sigma-Aldrich F6886) and pipette up and down until the powder is fully resuspended. Transfer to a 15-mL conical tube and add an additional 11 mL of DMSO to achieve a final concentration of 4.2 mg/mL. Store in 20- and 80-µL aliquots at −20°C.

Insulin Stock (0.5 mg/mL)

To 20 mL of sterile water, add 10 mg of insulin (Sigma-Aldrich I6634) and 100 µL of 1.0 N HCl. Mix well. Filter through a 0.22-µm filter. Store at 4°C for 4–6 wk.

Low-Ovomucoid Stock Solution (10×)

To prepare, add 3 g of BSA (Sigma-Aldrich A8806) to 150 mL D-PBS. Mix well. Add 3 g of trypsin inhibitor (Worthington LS003086) and mix to dissolve. Add ~1 mL of 1 N NaOH to adjust the pH to 7.4. Bring the volume to 200 mL with D-PBS. Filter-sterilize through a 0.22-µm filter. Make 1.0-mL aliquots and store at −20°C.

NAC Stock (5 mg/mL)

To prepare, dissolve 50 mg of *N*-acetyl-L-cysteine (NAC) powder (Sigma-Aldrich A8199) in 10 mL of Neurobasal Medium (Gibco/Life Technologies 21103). (The solution will be yellowish.) Filter through a 0.22-µm filter. Prepare 20- and 80-µL aliquots and store them at 4°C.

Schwann Cell Growth Medium

Reagent	Amount to add
80% DMEM/20% dH_20	18.5 mL
Penicillin/streptomycin (100×) (Gibco 15140-122)	200 µL
Insulin stock (0.5 mg/mL) <R>	200 µL
Sodium pyruvate (100 mM) (Gibco 11360-070)	200 µL
L-Glutamine (200 mM) (Gibco 25030-081)	200 µL
SATO supplement, NB-based (100×) <R>	200 µL
Thyroxine (T3) stock (4 µg/mL) <R>	200 µL
B-27 (Invitrogen) or N21-MAX (R&D Systems)	400 µL

Filter through a 0.22-µm filter and store for up to 1 wk at 4°C. Just before use, add the following:

NAC stock (5 mg/mL) <R>	20 µL
Forskolin stock (4.2 mg/mL) <R>	20 µL

SATO Supplement, NB-Based (100×)

1. Prepare the following stock solutions (these should be made fresh; do not reuse).
   - Combine 2.5 mg of progesterone (Sigma-Aldrich P8783) and 100 µL of ethanol to make a progesterone stock solution.
   - Combine 4.0 mg of sodium selenite (Sigma-Aldrich S5261), 10 µL of 1 N NaOH, and 10 mL of Neurobasal (NB; Gibco 21103-049) to make a sodium selenite stock solution.

 Cite this protocol as *Cold Spring Harb Protoc*; doi:10.1101/pdb.prot074989

2. Add the following to 80 mL of Neurobasal medium:

Reagent	Quantity	Final concentration in medium (1×)
BSA (Sigma-Aldrich A4161)	800 mg	100 µg/mL
Transferrin (Sigma-Aldrich T1147)	800 mg	100 µg/mL
Putrescine dihydrochloride (Sigma-Aldrich P5780)	128 mg	16 µg /mL
Progesterone stock solution	20 µL	60 ng/mL (0.2 µM)
Sodium selenite stock solution	800 µL	40 ng/mL

3. Mix well, and filter-sterilize through a prerinsed 0.22-µm filter. Make 200-µL or 800-µL aliquots, and store at −20°C.

Thyroxine (T3) Stock (4 µg/mL)

Dissolve 3.2 mg of 3,3′,5-triiodo-L-thyronine sodium salt (T3; Sigma-Aldrich T6397) in 400 µL of 0.1 N NaOH. Add 10 µL of T3 solution to 20 mL of Dulbecco's phosphate-buffered saline (D-PBS; Gibco 14287). Filter through a 0.22-µm filter, discarding the first 10 mL. Make 200-µL aliquots and store at −20°C.

ACKNOWLEDGMENTS

Many thanks to Lu Zhou for technical assistance and insight, Michio Painter for help with the protocol for dissociation, and Brad Zuchero for advice on cell culture conditions. This work was performed in Ben A. Barres' lab at Stanford University. His work was supported by the NIH (National Institutes of Health) R01 EY11310 (B.A.B.), the Adelson Medical Foundation (B.A.B), and the Christopher & Dana Reeve Foundation (B.A.B).

REFERENCES

Bansal R, Warrington AE, Gard AL, Ranscht B, Pfeiffer SE. 1989. Multiple and novel specificities of monoclonal antibodies O1, O4, and R-mAb used in the analysis of oligodendrocyte development. *J Neurosci Res* **24:** 548–557.

Lutz A. 2014. Purification of Schwann cells. *Cold Spring Harb Protoc* doi: 10.1101/pdb.top073981.

APPENDIX 1

Designing and Troubleshooting Immunopanning Protocols for Purifying Neural Cells

Ben A. Barres[1]

Stanford University School of Medicine, Stanford, California 94305

Purifying and culturing cells from the central nervous system (CNS) has proved to be an incredibly powerful tool for dissecting fundamental neuron and glial properties, and especially powerful in understanding neuronal–glial interactions. In a series of detailed protocols, we have provided step-by-step instructions for purifying and culturing specific types of neurons, glia, and vascular cells from the CNS by immunopanning. This chapter discusses common pitfalls and errors as well as important design considerations for the immunopanning procedure.

BACKGROUND

The nervous system is the most complex organ in our body. In humans, it has been estimated to contain as many as one trillion cells, including neurons, glia, and vascular cells. Understanding the interactions between these cells and their networks is critical for understanding the development and function of our nervous system, as well as how these processes go awry in disease. How can we begin to understand and unravel such complexity? In other organs, such as the immune system, purification of cell types has provided a powerful way forward.

As a PhD student with Linda Chun and David Corey in the 1980s, I developed a method to purify retinal ganglion cells (RGCs) from the postnatal rat retina using immunopanning (Barres et al. 1988). In immunopanning, antibodies are adsorbed to the surface of a Petri dish, and cell suspensions are incubated on this dish to select for or deplete cell types of interest. In a typical immunopanning purification, a cell suspension from the retina or other nervous system region is prepared and then passed over several immunopanning dishes consecutively, with the first dish or two being used to deplete unwanted cell types, such as microglia, and the final dish being used to select the cell type of interest. In my own laboratory over the past 20 years, we have continued to use immunopanning to develop purification and culture methods for many other CNS cell types, including other neuronal and glial types as well as vascular cells. The talented young people who have used and developed these methods, all of whom are now in their own laboratories, have each authored protocols detailing the purification and culture methods for these cell types (see Related Information).

Using the methods described in these protocols, it is now possible to prepare serum-free cultures of prospectively isolated neurons and glia whose properties in culture very closely resemble their in vivo counterparts. This provides the opportunity to begin to do many types of studies to understand the fundamental properties of neurons and glia, as well as their interactions with each other and with

[1]Correspondence: barres@stanford.edu

vascular cells. Below, troubleshooting information for immunopanning purification is provided, and considerations for the design of successful immunopanning protocols are discussed.

COMMON ERRORS IN IMMUNOPANNING

Purification of cells by panning is simple and provides high-yield, high-viability cells reliably, once you get the hang of it. Generally, the method takes practice, as every step must be performed correctly to have the cells still in good shape by the end of the procedure. Two of the most common errors that are made are (1) not using Worthington Biochemical as the source for papain and (2) not using appropriately prepared and stored trypsin for the removal of the purified cells from the final panning dish. (Trypsin will cleave itself over time if not stored at −80°C—though genetically engineered preparations of trypsin are now available that are modified to prevent this.)

Panning is trivial, involving only three steps: (1) enzymatic preparation of a cell suspension, (2) passing this suspension over a series of antibody-coated dishes, and (3) removing the purified cells from the final dish. In general, the following guidelines should be followed: attempt to minimize the length of the entire procedure, avoid cooling mammalian neurons lower than room temperature even for only a few minutes, and never centrifuge enzymatically isolated neural cells in a solution containing less than 10% fetal calf serum (FCS) or 0.5% bovine serum albumin (BSA). This is particularly true if the cells have been twice exposed to enzyme—for instance, to papain for preparation of the initial cell suspension followed by trypsin to remove them from the dish. Centrifugation without sufficient protein in the solution after the trypsin step will cause death of most of the cells. In addition, avoid using bicarbonate-buffered solutions in the procedures, except with cells being stored in a CO_2 incubator.

LOW YIELD

When a low yield of cells is obtained from an immunopanning purification, it is critical to determine at what stage of the procedure the yield is low. If it is low after the initial preparation of the cell suspension, this suggests that the papain or DNase may not be active or used at sufficient concentration (or that magnesium has been omitted from the DNase solution, which needs to contain at least 100 μM Mg^{2+}). On the other hand, if the yield is normal after preparation of the cell suspension but the density of cells looks low on the final panning dish, this suggests that the problem was with the preparation of that final dish. Essential questions would be was the secondary antibody (the initial antibody placed on the dish) the right isotype, was it applied in a buffer at pH 9.5, is the dish a Petri dish and not a tissue culture dish, and was the primary antibody used at too high or too low a concentration? If the density of cells looks good on the final dish, but the final yield after the cells have been trypsinized is low, then it may be that the trypsin was not sufficiently active, or the cells were not centrifuged in a solution containing sufficient protein concentration. In addition, temperature control is critical for good viability. As has already been mentioned, chilling of mammalian neurons, even briefly, can lead to very low viability. Another, easily overlooked problem is that old hot water baths often become unstable and undergo constant temperature oscillations that can lead to incubation temperatures that are too high. Even half a degree or one degree too high can lead to low viability and cell death.

If the viability of the purified cells is low, it is similarly critical to determine when the viability became low. If the viability is <85% after preparation of the tissue cell suspension (before panning), typically very few of these cells will survive in culture. If the viability is low immediately after removing the cells from the final panning dish, this suggests either that the trypsin was not active or that the cell surface antigen was not trypsin-sensitive; typically, the yield would be very low in these cases as well. If the yield and viability are fine after panning but viability is poor after a few days of culture, then there is almost certainly a problem with the culture conditions: The medium or growth factors could be expired or defective, the substrate may not be optimal, the glass coverslips might not have been cleaned sufficiently, or the tissue culture plastic may be leaching a toxic plastic stabilizing resin (these resins can also leak from certain brands of plastic syringes). A very common problem is that the CO_2 concentration in the incubator is not correct, even though the incubator CO_2 meter claims that it is. Incubators need to be calibrated frequently.

CULTURE MEDIA

The composition of serum-free media is critical to good viability. In addition to having appropriate growth factors or other signals necessary to prevent apoptosis, medium must be the correct pH and osmolarity. Common base medium solutions used for culture of neural cells vary dramatically in their osmolarity. For instance, Dulbecco's modified Eagle's medium (DMEM) has an osmolarity of ~330 mOsm, whereas Neurobasal has an osmolarity of only ~220. (Normal cerebrospinal fluid is more typically ~290–300 mOsm.) It is important to keep in mind that Neurobasal is an extremely minimal medium, with fewer substances in lower concentrations (such as amino acids) when compared to DMEM. In our experience, it is therefore much easier for Neurobasal to become inactive. To avoid this potential problem, we sometimes use DMEM as the base medium, but dilute it by ~25% to bring it down to the osmolarity of Neurobasal. Alternatively, DMEM and Neurobasal can be used as a 50:50 mixture, which also brings osmolarity closer to normal. The osmolarity of the final medium is increased, of course, by the addition of the many additives such as glutamine, penicillin/streptomycin, etc.—often by as much as another 35 mOsm.

Another very important consideration in achieving good viability is the quality of the serum-free additives that are included in the final culture medium, such as the Bottenstein–Sato serum-free additive, B-27, or NS21. In our experience, commercially obtained B-27 (which is tested by the company only for hippocampal neurons) has been associated in recent years with toxicity for at least some or all of the cell types described in our protocols. For this reason, we have moved to the NS21 additive, which we formulate ourselves (Chen et al. 2008). The Bottenstein–Sato reagent is made differently depending upon whether it is commercially obtained or homemade. The reagents in this additive typically include transferrin, sodium selenite, putrescine, progesterone, BSA, and, optionally, thyroxine and tri-iodothyronine. In our experience, the transferrin and selenite are the most critical, as they are crucial for both survival and proliferation of cells. Because transferrin is expensive, commercial sources typically have far lower concentrations than are optimal. When using BSA with primary cells, it is important to use a crystalline, relatively pure preparation to avoid toxic contaminants.

TIPS FOR DESIGNING NEW PANNING PROTOCOLS

General points that may be helpful in troubleshooting and designing new immunopanning protocols are described in the following section. In addition, a list of immunolabeling antibodies useful for identifying cell types is provided in Table 1.

Enzymatic Preparation

For panning, it is preferable to use antibodies that recognize surface antigens that are not cleaved off by the papain or trypsin used to prepare the cell suspension. For this reason, antibodies to glycolipids (e.g., GC) or gangliosides (e.g., A2B5) are ideal. In some cases when the antigen is a protein, the enzymes used do not completely cleave the antigen of interest and so enzymes can still be used (e.g., RAN-2 and Thy1). In many cases, however, the enzymes completely destroy the antigen. It is still possible to pan using antibodies to this antigen, if you allow the cell suspension to recover at 37°C for 30 min to 1 h. This allows protein already in the secretory pathway to be transferred to the cell membrane. If you do this, the cell suspension should be pipetted and passed through a nylon mesh filter (Nitex) before it is added to the panning dish, so that any cell clumps that have aggregated during

TABLE 1. Immunolabeling antibodies for identifying cell types

Cell type	Antigen	Antibody source	Dilution
Astrocytes	GFAP (anti-glial fibrillary acidic protein)	Sigma-Aldrich G3893	1:2000
	Anti-AQP4 (aquaporin 4)	Sigma-Aldrich A5971	1:1000
	GLT-1 (glutamate transporter 1; EAAT2)	Abcam ab41621	1:5000
Endocytes	CD31 (PECAM-1)	BD Pharmingen 553370	1:500
	Occludin	Invitrogen 71–1500	1:250
Pericytes	NG2 (chondroitin sulfate proteoglycan)	Millipore AB5320	1:400
	PDGFRβ	R&D Systems AF385, AF1042	1:500
OPC	NG2	Millipore AB5320	1:400
	CNPase (2′,3′-cyclic nucleotide 3′-phosphodiesterase)	Millipore MAB326	1:100
	PDGFR α	BD Pharmingen 558774	1:5000
	O4	Millipore MAB345	1:100
OL	MBP (myelin basic protein)	Abcam ab7349	1:100
	PLP (proteolipid protein)	Millipore MAB388	1:400
	MAG (myelin associated glycoprotein)	Millipore MAB1567	1:50

the incubation period are eliminated. Clumps are a major threat to purity when they adhere to the final panning dish, as they can bring down unwanted contaminating cell types.

Panning

Always pan at room temperature. Never pan at 37°C, because this allows activation of cell adhesion mechanisms and lets unwanted contaminating cell types adhere to panning dishes. Always use at least two panning dishes. The first should be coated with an antibody to deplete an unwanted contaminating cell type or with an irrelevant antibody to ensure that microglia/macrophages are depleted. Microglia/macrophages represent a high percentage of cells in the developing nervous system (e.g., in the developing optic nerve, they make up ~6% of cells) and stick via their Fc receptors to the first antibody-coated dish they contact. They will not be successfully eliminated by panning on a dish that is not antibody-coated. An alternative for removing microglia/macrophages is to just coat the first dish with the lectin BSLI (*Bandeiraea simplicifolia* Lectin I; this lectin also binds to brain endothelial cells) or the CD45 monoclonal antibody.

Regarding the antibodies used to coat panning dishes, the first antibody (the goat anti-rabbit or anti-mouse) should always be affinity purified. That is, the main protein in this solution should be the antibody of interest; otherwise, irrelevant proteins such as BSA will hog most of the binding sites on the Petri dishes. This is why monoclonal supernatants, for instance, which generally contain 10% FCS, cannot be put straight onto the dish without precoating with an affinity purified anti-mouse antibody. If you happen to have an affinity-purified antibody (or a lectin) to the cell surface antigen of interest, then it can be put straight onto the dish (in a pH 9.5 solution, which is critical for good absorbance of the protein to the plastic).

In addition, the second antibody (the one that actually binds to the cell surface antigen of interest; see Table 2) should be a monoclonal supernatant whenever possible. The reason for this is that the only mouse antibody type in the supernatant is the one of interest. If you use a polyclonal antiserum, then there will be many irrelevant antibodies that will hog (anti-rabbit) sites on the panning dish. Similarly, many antibodies secreted into ascites are irrelevant host-derived antibodies and will also hog (anti-mouse) sites on the panning dish. In some cases, ascites from IgM-secreting hybridoma cells can be used as long as an anti-mouse IgM, μ-chain-specific antibody is used to first coat the dish, as there do not seem to be many irrelevant IgM antibodies in ascites.

If there is no alternative and a polyclonal antiserum or ascites must be used, then it is still possible to use this antibody for panning. In this case, however, the panning dish must be coated with only one antibody (the anti-rabbit or anti-mouse antibody) and the cell suspension should be incubated with the rabbit or mouse antibody directly. In this way, the wanted antibodies will bind to the cell surface and the unwanted ones can be eliminated by centrifuging the cell suspension and discarding the supernatant (and preferably washing the suspension one more time).

TABLE 2. Primary antibodies for immunopanning purification of neural cells

Antibody	Source
Mouse anti-p75 NGF receptor antibody	Abcam ab6172
Mouse anti-CTB (clone 3D11)	Meridian Life Science C86204M
Rat anti-mouse CD140a (PDGFRα)	BD Pharmingen 558774
Mouse anti-rat CD31	Fitzgerald 10R-CD31gRT
Mouse anti-rat CD45	AbD Serotec MCA589
Rat anti-mouse CD31	BD Pharmingen 553370
Rat anti-mouse CD45	BD Pharmingen Rat anti-ms550539 (Clone 30F-11)
C5 hybridoma supernatant	Mouse
Goat anti-human PDGFRβ	R&D Systems AF385
Goat-anti mouse PDGFRβ	R&D Systems AF1042
T11D7 supernatant	Mouse
Rabbit anti-rat antimacrophage sera	Accurate Chemical AIAD51240
Ran-2 hybridoma supernatant	Mouse
Anti-O1	Millipore MAB344
Anti-O4	Millipore MAB345

Removing Cells from the Final Dish

The most delicate step in panning is removing the tightly adhering purified cells from the panning dish. Although saturating amount of antibody can be used on the first panning dishes for depletion of contaminants from the cell suspension, the concentration of the antibody on the final panning dish is crucial and must be titrated carefully. If the cells are too tightly bound they will be difficult to release viably; if they are too loosely bound they will wash off and the yield will be low. This is particularly crucial if the antigen is not destroyed by trypsin, as antibodies are not well cleaved by trypsin. In general, an optimal concentration of antibody causes about half of the adherent cells to be phase dark and flat (tightly adherent) and the other half to be phase bright and roundish (less tightly adherent). This can be determined by a titration experiment in which the yield of viable cells is optimized; typically, supernatants on the final dish are diluted 1:20 and ascites ~1:2000.

In general, if the antigen used to purify cells on the final panning dish is not trypsin-sensitive, then the antibody used needs to be an IgM rather than an IgG. The IgG-bound cells are difficult to release viably, perhaps because of high affinity for the antigen. If the only antibody available is an IgG and the antigen is not trypsin-sensitive, it is still possible to purify the cells. In this case, the final dish must be coated with either protein A or protein G (or a mixture of both) rather than an anti-mouse IgG antibody. Protein A and protein G are easily cleaved by trypsin and the cells are easily released.

Keeping the above considerations in mind, it should be straightforward to achieve purified cells in high yield with excellent viability and to design new immunopanning procedures.

RELATED INFORMATION

Step-by-step instructions for purifying and culturing specific types of neurons, glia, and vascular cells from the central nervous system by immunopanning are provided in the following protocols: Chapter 8, Protocol 1: Purification of Oligodendrocyte Precursor Cells from Rat Cortices by Immunopanning (Dugas and Emery); Chapter 8, Protocol 2: Purification of Oligodendrocyte Precursor Cells from Mouse Cortices by Immunopanning (Emery and Dugas); Chapter 3, Protocol 1: Purification and Culture of Spinal Motor Neurons from Rat Embryos (Graber and Harris); Chapter 5, Protocol 1: Purification of Rat and Mouse Astrocytes by Immunopanning (Foo); Chapter 1, Protocol 1: Purification and Culture of Retinal Ganglion Cells from Rodents (Winzeler and Wang); Chapter 2, Protocol 2: Immunopanning of Retrograde-Labeled Corticospinal Motor Neurons from Early Postnatal Rodents (Mandemakers); Chapter 4, Protocol 1: Purification of Dorsal Root Ganglion Neurons from Rat by Immunopanning (Zuchero); Chapter 6, Protocol 1: Purification of Pericytes from Rodent Optic

Nerve by Immunopanning (Zhou et al. [a]); Chapter 7, Protocol 1: Purification of Endothelial Cells from Rodent Brain by Immunopanning (Zhou et al. [b]); Chapter 9, Protocol 1: Myelinating Cocultures of Rat Retinal Ganglion Cell Reaggregates and Optic Nerve Oligodendrocyte Precursor Cells (Watkins and Scholze); and Chapter 10, Protocol 1: Purification of Schwann Cells from the Neonatal and Injured Adult Mouse Peripheral Nerve (Lutz).

REFERENCES

Barres BA, Silverstein BE, Corey DP, Chun LL. 1988. Immunological, morphological, and electrophysiological variation among retinal ganglion cells purified by panning. *Neuron* **1:** 791–803.

Chen Y, Stevens B, Chang J, Milbrandt J, Barres BA, Hell JW. 2008. NS21: Re-defined and modified supplement B27 for neuronal cultures. *J Neurosci Meth* **171:** 239–247.

Dugas JC, Emery B. 2013. Purification of oligodendrocyte precursor cells from rat cortices by immunopanning. *Cold Spring Harb Protoc* doi: 10.1101/pdb.prot070862.

Emery B, Dugas JC. 2013. Purification of oligodendrocyte precursor cells from mouse cortices by immunopanning. *Cold Spring Harb Protoc* doi: 10.1101/pdb.prot073973.

Foo LC. 2013. Purification of rat and mouse astrocytes by immunopanning. *Cold Spring Harb Protoc* doi: 10.1101/pdb.prot074211.

Graber DJ, Harris BT. 2013. Purification and culture of spinal motor neurons from rat embryos. *Cold Spring Harb Protoc* doi: 10.1101/pdb.prot074161.

Lutz AB. 2014. Purification of Schwann cells from the neonatal and injured adult mouse peripheral nerve. *Cold Spring Harb Protoc* doi: 10.1101/pdb.prot074989.

Mandemakers W. 2014. Immunopanning of retrograde-labeled corticospinal motor neurons from early postnatal rodents. *Cold Spring Harb Protoc* doi: 10.1101/pdb.prot074930.

Watkins TA, Scholze AR. 2014. Myelinating cocultures of rat retinal ganglion cell reaggregates and optic nerve oligodendrocyte precursor cells. *Cold Spring Harb Protoc* doi: 10.1101/pdb.prot074971.

Winzeler A, Wang JT. 2013. Purification and culture of retinal ganglion cells from rodents. *Cold Spring Harb Protoc* doi: 10.1101/pdb.prot074906.

Zhou L, Sohet F, Daneman R. 2014a. Purification of pericytes from rodent optic nerve by immunopanning. *Cold Spring Harb Protoc* doi: 10.1101/pdb.prot074955.

Zhou L, Sohet F, Daneman R. 2014b. Purification of endothelial cells from rodent brain by immunopanning. *Cold Spring Harb Protoc* doi: 10.1101/pdb.prot074963.

Zuchero JB. 2014. Purification of dorsal root ganglion neurons from rat by immunopanning. *Cold Spring Harb Protoc* doi: 10.1101/pdb.prot074948.

APPENDIX 2

General Safety and Hazardous Material Information

Users should always consult individual manufacturers, the manufacturers' safety guidelines and other resources, including local safety offices, for current and specific product information and for guidance regarding the use and disposal of hazardous materials.

PRIMARY SAFETY INFORMATION RESOURCES FOR LABORATORY PERSONNEL

Institutional Safety Office. The best source of toxicity, hazard, storage, and disposal information is your institutional safety office, which maintains and makes available the most current information. Always consult this office for proper use and disposal procedures.

Post the phone numbers for your local safety office, security office, poison control center, and laboratory emergency personnel in an obvious place in your laboratory.

Material Safety Data Sheets (MSDSs). The Occupational Safety and Health Administration (OSHA) requires that MSDSs accompany all hazardous products that are shipped. These data sheets contain detailed safety information. MSDSs should be filed in the laboratory in a central location as a reference guide.

GENERAL SAFETY AND DISPOSAL CAUTIONS

The guidance offered here is intended to be generally applicable. However, proper waste disposal procedures vary among institutions; therefore, always consult your local safety office for specific instructions. All chemically constituted waste must be disposed of in a suitable container clearly labeled with the type of material it contains and the date the waste was initiated.

It is essential for laboratory workers to be familiar with the potential hazards of materials used in laboratory experiments and to follow recommended procedures for their use, handling, storage, and disposal.

The following general cautions should always be observed.

- **Before beginning the procedure,** become completely familiar with the properties of substances to be used.
- **The absence of a warning** does not necessarily mean that the material is safe, because information may not always be complete or available.
- **If exposed** to toxic substances, contact your local safety office immediately for instructions.
- **Use proper disposal procedures** for all chemical, biological, and radioactive waste.
- **For specific guidelines on appropriate gloves to use,** consult your local safety office.
- **Handle concentrated acids and bases** with great care. Wear goggles and appropriate gloves. A face shield should be worn when handling large quantities.

  Do not mix strong acids with organic solvents because they may react. Sulfuric acid and nitric acid especially may react highly exothermically and cause fires and explosions.

  Do not mix strong bases with halogenated solvents because they may form reactive carbenes that can lead to explosions.
- **Handle and store pressurized gas containers** with caution because they may contain flammable, toxic, or corrosive gases; asphyxiants; or oxidizers. For proper procedures, consult the Material Safety Data Sheet that is required to be provided by your vendor.
- **Never pipette** solutions using mouth suction. This method is not sterile and can be dangerous. Always use a pipette aid or bulb.
- **Keep halogenated and nonhalogenated** solvents separately (e.g., mixing chloroform and acetone can cause unexpected reactions in the presence of bases). Halogenated solvents are organic solvents such as chloroform, dichloromethane, trichlorotrifluoroethane, and dichloroethane. Nonhalogenated solvents include pentane, heptane, ethanol, methanol, benzene, toluene, *N,N*-dimethylformamide (DMF), dimethylsulfoxide (DMSO), and acetonitrile.
- **Laser radiation,** visible or invisible, can cause severe damage to the eyes and skin. Take proper precautions to prevent exposure to direct and reflected beams. Always follow the manufacturer's safety guidelines and consult your local safety office. See caution below for more detailed information.
- **Flash lamps,** because of their light intensity, can be harmful to the eyes. They also may explode on occasion. Wear appropriate eye protection and follow the manufacturer's guidelines.
- **Photographic fixatives, developers, and photoresists** also contain chemicals that can be harmful. Handle them with care and follow the manufacturer's directions.
- **Power supplies and electrophoresis equipment** pose serious fire hazard and electrical shock hazards if not used properly.
- **Microwave ovens and autoclaves** in the laboratory require certain precautions. Accidents have occurred involving their use (e.g., when melting agar or Bacto Agar stored in bottles or when sterilizing). If the screw top is not completely removed and there is inadequate space for the steam to vent, the bottles can explode and cause severe injury when the containers are removed from the microwave or autoclave. Always completely remove bottle caps before microwaving or autoclaving. An alternative method for routine agarose gels that do not require sterile agar is to weigh out the agar and place the solution in a flask.
- **Ultrasonicators** use high-frequency sound waves (16–100 kHz) for cell disruption and other purposes. This "ultrasound," conducted through air, does not pose a direct hazard to humans, but the associated high volumes of audible sound can cause a variety of effects, including headache, nausea, and tinnitus. Direct contact of the body with high-intensity ultrasound (not medical

imaging equipment) should be avoided. Use appropriate ear protection and display signs on the door(s) of laboratories where the units are used.

- **Use extreme caution when handling cutting devices,** such as microtome blades, scalpels, razor blades, or needles. Microtome blades are extremely sharp! Use care when sectioning. If unfamiliar with their use, have an experienced user demonstrate proper procedures. For proper disposal, use the "sharps" disposal container in your laboratory. Discard used needles *unshielded*, with the syringe still attached. This prevents injuries and possible infections when manipulating used needles because many accidents occur while trying to replace the needle shield. Injuries may also be caused by broken Pasteur pipettes, coverslips, or slides.
- **Procedures for the humane treatment of animals** must be observed at all times. Consult your local animal facility for guidelines. Animals, such as rats, are known to induce allergies that can increase in intensity with repeated exposure. Always wear a lab coat and gloves when handling these animals. If allergies to dander or saliva are known, wear a mask.

DISPOSAL OF LABORATORY WASTE

There are specific regulatory requirements for the disposal of all medical waste and biological samples mandated by the U.S. Environmental Protection Agency (see http://www.epa.gov/epawaste/hazard/tsd/index.htm) and regulated by the individual states and territories (see http://www.epa.gov/epawaste/wyl/stateprograms.htm). Medical and biological samples that require special handling and disposal are generally termed Medical Pathological Waste (MPW), and medical, veterinary, and biological facilities will have programs for the collection of MPW and its disposal. Restrictions on how radioactive waste can be disposed of as regulated by the U.S. Nuclear Regulatory Commission can be found in 10 CFR 20.2001, General requirements for waste disposal (see http://www.nrc.gov/reading-rm/doc-collections/cfr/part020/part020-2001.html) or the individual Agreement States. The preferred method for the disposal of radioactively contaminated MPW'is decay-in-storage (see http://www.nrc.gov/reading-rm/doc-collections/cfr/part035/part035-0092.html).

Waste and any materials contaminated with biohazardous materials must be decontaminated and disposed of as regulated medical waste. No harmful substances should be released into the environment in an uncontrolled manner. This includes all tissue samples, needles, syringes, scalpels, etc. Be sure to contact your institution's safety office concerning the proper practices associated with the handling and disposal of biohazardous waste.

Some basic rules are outlined below. For treatment of radioactive and biological waste, see sections on Radioactive Safety Procedures and Biological Safety Procedures.

- In practice, only **neutral aqueous solutions** without heavy metal ions and without organic solvents can be poured down the drain (e.g., most buffers). Acid and basic aqueous solutions need to be neutralized cautiously before their disposal by this method.
- For proper disposal of **strong acids and bases**, dilute them by placing the acid or base onto ice and neutralize them. Do not pour water into them. If the solution does not contain any other toxic compound, the salts can be flushed down the drain.
- For disposal of **other liquid waste**, similar chemicals can be collected and disposed of together, whereas chemically different wastes should be collected separately. This avoids chemical reactions between components of the mixture (see above). Collect at least inorganic aqueous waste, non-halogenated solvents, and halogenated solvents separately.
- Waste **from photo processing and automatic developers** should be collected separately to recycle the silver traces found in it.

RADIOACTIVE SAFETY PROCEDURES

In the United States and other countries, the access to radioactive substances is strictly controlled. You may be required to become a registered user (e.g., by attending a mandatory seminar and receiving a personal dosimeter). A convenient calculator to perform routine radioactivity calculations can also be found at http://www.graphpad.com/quickcalcs/ChemMenu.cfm.

If you have never worked with radioactivity before, follow the steps below.

- ***Try to avoid it!*** Many experiments that are traditionally performed with the help of radioactivity can now be done using alternatives based on fluorescence or chemiluminescence and colorimetric assays, including, for example, DNA sequencing, Southern and northern blots, and protein kinase assays. However, in other cases (e.g., metabolic labeling of cells), use of radioactivity cannot be avoided.
- **Be informed.** While planning an experiment that involves the use of radioactivity, include the physicochemical properties of the isotope (half-life, emission type, and energy), the chemical form of the radioactivity, its radioactive concentration (specific activity), total amount, and its chemical concentration. Order and use only as much as is really needed.
- **Familiarize yourself** with the designated working area. Perform a mental and practical dry run (replacing radioactivity with a colored solution) to make sure that all equipment needed is available and to get used to working behind a shield. Handle your samples as if sterility would be required to avoid contamination.
- **Always wear appropriate gloves**, lab coat, and safety goggles when handling radioactive material.
- **Check the work area** for contamination before, during, and after your experiment (including your lab coat, hands, and shoes).
- **Localize your radioactivity.** Avoid formation of aerosols or contamination of large volumes of buffers.
- **Liquid scintillation cocktails** are often used to quantitate radioactivity. They contain organic solvents and small amounts of organic compounds. Try to avoid contact with the skin. After use, they should be regarded as radioactive waste; the filled vials are usually collected in designated containers, separate from other (aqueous) liquid radioactive waste.
- **Dispose of radioactive waste** only into designated, shielded containers (separated by isotope, physical form [dry/liquid], and chemical form [aqueous/organic solvent phase]). Always consult your safety office for further guidance in the appropriate disposal of radioactive materials.
- Among the experiments requiring **special precautions** are those that use [^{35}S]methionine and ^{125}I, because of the dangers of airborne radioactivity. [^{35}S]methionine decomposes during storage into sulfoxide gases, which are released when the vial is opened. The isotope ^{125}I accumulates in the thyroid and is a potential health hazard. ^{125}I is used for the preparation of Bolton–Hunter reagent to radioiodinate proteins. Consult your local safety office for further guidance in the appropriate use and disposal of these radioactive materials before initiating any experiments. Wear appropriate gloves when handling potentially volatile radioactive substances, and work only in a radioiodine fume hood.

BIOLOGICAL SAFETY PROCEDURES

Biological safety fulfills three purposes: to avoid contamination of your biological sample with other species; to avoid exposure of the researcher to the sample; and to avoid release of living material into the environment. Biological safety begins with the receipt of the living sample; continues with its storage, handling, and propagation; and ends only with the proper disposal of all contaminated materials. A catalog of operations known as "sterile handling" is usually employed in manipulating living matter. However, the actual manner of treatment largely depends on the actual sample, which can be quite diverse: *Escherichia coli* and other bacterial strains, yeasts, tissues of animal or plant origin, cultures of mammalian cells, or even derivatives from human blood are routinely handled in a biological laboratory. Two of these, bacteria and human blood products, are discussed in more detail below.

The Department of Health, Education, and Welfare (HEW) has classified various bacteria into different categories with regard to shipping requirements (see Sanderson and Zeigler 1991). Non-pathogenic strains of *E.coli* (such as K12) and *Bacillus subtilis* are in Class 1 and are considered to present no or minimal hazard under normal shipping conditions. However, *Salmonella*, *Haemophilus*, and certain strains of *Streptomyces* and *Pseudomonas* are in Class 2. Class 2 bacteria are "[a]gents of ordinary potential hazard: agents which produce disease of varying degrees of severity... but which are contained by ordinary laboratory techniques." Contact your institution's safety office concerning shipping biological material.

Human blood, blood products, and tissues may contain occult infectious materials such as hepatitis B virus and human immunodeficiency virus (HIV) that may result in laboratory-acquired infections. Investigators working with lymphoblast cell lines transformed by Epstein–Barr virus (EBV) are also at risk of EBV infection. Any human blood, blood products, or tissues should be considered a biohazard and should be handled accordingly until proved otherwise. Wear appropriate disposable gloves, use mechanical pipetting devices, work in a biological safety cabinet, protect against the possibility of aerosol generation, and disinfect all waste materials before disposal. Autoclave contaminated plasticware before disposal; autoclave contaminated liquids or treat with bleach (10% [v/v] final concentration) for at least 30 minutes before disposal (this is valid also for used bacterial media).

Always consult your local institutional safety officer for specific handling and disposal procedures of your samples. Further information can be found in the Frequently Asked Questions of the ATCC homepage (http://www.atcc.org) and is also available from the National Institute of Environmental Health and Human Services, Biological Safety (http://www.niehs.nih.gov/about/stewardship).

GENERAL PROPERTIES OF COMMON HAZARDOUS CHEMICALS

The hazardous materials list can be summarized in the following categories.

- **Inorganic acids**, such as hydrochloric, sulfuric, nitric, or phosphoric, are colorless liquids with stinging vapors. Avoid spills on skin or clothing. Spills should be diluted with large amounts of water. The concentrated forms of these acids can destroy paper, textiles, and skin and cause serious injury to the eyes.
- **Inorganic bases**, such as sodium hydroxide, are white solids that dissolve in water and under heat development. Concentrated solutions will slowly dissolve skin and even fingernails.
- **Salts of heavy metals** are usually colored, powdered solids that dissolve in water. Many of them are potent enzyme inhibitors and therefore toxic to humans and the environment (e.g., fish and algae).
- Most **organic solvents** are flammable volatile liquids. Avoid breathing the vapors, which can cause nausea or dizziness. Also avoid skin contact.
- **Other organic compounds** including organosulfur compounds, such as mercaptoethanol or organic amines, can have very unpleasant odors. Others are highly reactive and should be handled with appropriate care.
- If improperly handled, **dyes and their solutions** can stain not only your sample but also your skin and clothing. Some are also mutagenic (e.g., ethidium bromide), carcinogenic, and toxic.

- **Nearly all names ending with "ase"** (e.g., catalase, β-glucuronidase, or zymolyase) refer to enzymes. There are also other enzymes with nonsystematic names such as pepsin. Many of them are provided by manufacturers in preparations containing buffering substances, etc. Be aware of the individual properties of materials contained in these substances.
- **Toxic compounds** are often used to manipulate cells. They can be dangerous and should be handled appropriately.
- Be aware that several of the compounds listed have not been thoroughly studied with respect to their toxicological properties. Handle each chemical with appropriate respect. Although the toxic effects of a compound can be quantified (e.g., LD_{50} values), this is not possible for carcinogens or mutagens where one single exposure can have an effect. Also realize that dangers related to a given compound may also depend on its physical state (fine powder vs. large crystals/diethyl ether vs. glycerol/dry ice vs. carbon dioxide under pressure in a gas bomb). Anticipate under which circumstances during an experiment exposure is most likely to occur and how best to protect yourself and your environment.

Cold Spring Harbor Laboratory Press (CSHLP) has used its best efforts in collecting and preparing the material contained herein but does not assume, and hereby disclaims, any liability for any loss or damage caused by errors and omissions in the publication, whether such errors and omissions result from negligence, accident, or any other cause. CSHLP does not assume responsibility for the user's failure to consult more complete information regarding the hazardous substances listed in this publication.

REFERENCE

Sanderson KE, Zeigler DR. 1991. Storing, shipping, and maintaining records on bacterial strains. *Methods Enzymol* **204**: 248–264.

WWW RESOURCES

ATCC Home page http://www.atcc.org

ATCC, for Sample Handling (in Frequently Asked Questions) http://www.atcc.org/CulturesandProducts/TechnicalSupport/FrequentlyAskedQuestions/tabid/469/Default.aspx

GraphPad Software, Radioactivity Calculations http://www.graphpad.com/quickcalcs/ChemMenu.cfm

National Institute of Environmental Health and Human Services, Biological Safety (NIEHS) http://www.niehs.nih.gov/about/stewardship

U.S. Environmental Protection Agency (EPA), Federal waste disposal regulations, Laboratory http://www.epa.gov/epawaste/hazard/tsd/index.htm

U.S. Environmental Protection Agency (EPA), Individual States and Territories http://www.epa.gov/epawaste/wyl/stateprograms.htm

U.S. Nuclear Regulatory Commission (NRC), Medical Pathological Radioactively Contaminated Waste (Decay-in-Storage) http://www.nrc.gov/reading-rm/doc-collections/cfr/part035/part035-0092.html

U.S. Nuclear Regulatory Commission (NRC), Radioactive Waste Disposal Regulations: General Requirements http://www.nrc.gov/reading-rm/doc-collections/cfr/part020/part020-2001.html

Index

Page references followed by f denote figures; those followed by t denote tables.

K

L

M

N

O

S

T

U

V

W